546.36. 50

PADS - P. 56

RESIDENTIAL FOUNDATIONS

RESIDENTIAL FOUNDATIONS

Design, Behavior and Repair

Second Edition

ROBERT WADE BROWN

VNR VAN NOSTRAND REINHOLD COMPANY
———————————— New York

Manufactured in the United States of America

Published by Van Nostrand Reinhold Company Inc.
135 West 50th Street
New York, New York 10020

Van Nostrand Reinhold Company Limited
Molly Millars Lane
Wokingham, Berkshire RG11 2PY, England

Van Nostrand Reinhold
480 Latrobe Street
Melbourne, Victoria 3000, Australia

Macmillan of Canada
Division of Gage Publishing Limited
164 Commander Boulevard
Agincourt, Ontario M1S 3C7, Canada

15 14 13 12 11 10 9 8 7 6 5 4 3 2

Library of Congress Cataloging in Publication Data

Brown, Robert Wade.
    Residential foundations.

    Includes bibliographical references and index.
    1. Foundations.   2. Dwellings.   I. Title.
TH2101.B73   1984   690'.11   83-10391
ISBN 0-442-21302-6

PREFACE

This book is the result of fifteen years of activity and research devoted to analyzing and correcting foundation failures. It is intended as a reference for engineers, architects, appraisers, realtors, lenders and, hopefully, even the more sophisticated home owners.

Presentation of the subject begins with a brief insight into the technical areas which influence foundation design and behavior and closes with a look at some methods for correcting and preventing failures. The selection of subject material was influenced by other authors and associates to whom I am grateful. The bibliography will enable the serious student to explore the subject in as much detail as might be desired.

The author is particularly grateful to my friend and colleague, Dr. Cecil Smith, of the Civil and Mechanical Engineering Department at Southern Methodist University for his review and comments. I also wish to express appreciation to Brown Foundation Repair & Consulting, Inc., for making the publication possible and to my family for their patience and support in preparing the text.

ROBERT WADE BROWN

INTRODUCTION

Foundations, like all structures, are designed against "failure." Failure, for purposes herein, may be defined as either "total collapse" or a condition exhibiting sufficient deflection from the original construction so as to require remedial attention. This latter condition, distress failure, provides the basis for this book.

Foundations designs against ultimate failure are predicated both on the original nature of the structure (size, loads, type construction, etc.) and on any later conditions under which the structure must exist. Two principal factors which influence both design and any subsequent failure are the properties of the particular bearing soil, compounded by the interaction of climatic conditions. Generally speaking, the stability of the soil is of prime importance, more particularly, the stability as affected by the tendency of the clay constituents for volumetric changes brought about by changes in moisture content.

In simpler terms, foundations constructed on highly active clay soils demonstrate a propensity for distress failures, particularly when located in areas of widely variant moisture. Unfortunately, many areas of the United States are susceptible to this coexistance — to one extent or another. The South, Midwest and Southwest represent the areas of highest incidence. It is interesting to note that the annual cost to Americans for foundation repairs exceeds $2 billion, according to a study reported by Jones and Holtz.*

*Jones, D. E., Jr. and Holtz, W. G., "Expansive Soils — The Hidden Disaster," *Civil Engineering, ASCE*, vol 43, Aug. 1973, pp. 49-51.

The primary purpose of this book is not to delve into the criteria of foundation design but to discuss foundation distress: the cause, the prevention, and the cure.

In many areas, the primary cause of foundation failures can be directly attributed to the clay mineral, montmorillonite (or smectite), present in the soil and changes in moisture therein. For example, a relatively pure sodium-substituted montmorillonite clay has the capacity to free swell up to twenty fold when taken from a dehydrated to saturated state. Other clays such as illite, attapulgite, and kaolinite also exhibit the swell potential, though to a lesser extent. In the course of this swell a confined clay could exert pressures of several tons per square foot. Hence, it becomes quite evident that changes in soil moisture contribute to substantial forces which ultimately could cause foundation deflections and "failure" — particularly in the case of residential construction.

When a foundation is constructed, it covers or "protects" an area of bearing soil. The causes for moisture imbalance within this confined area can then be those attributable to either variations in natural ambient conditions (rain, heat, wind) and/or to "man-created" conditions such as domestic water leaks.

Assuming proper drainage at the foundation perimeter, the ambient influences to soil moisture will primarily affect only the outer uncovered soil. The effects of this are generally manifested in failures attributed to settlement or shrinkage of the soil from a loss of water, and, accordingly, referred to as settlement.

Water that collects under the foundation, regardless of origin, is another problem. This water is confined, accumulative, and particularly serious — accounting for a high percentage of all failures. The source can be from either domestic leaks or nature. Foundation failures resulting from this excessive water under the foundation are classically termed upheaval.

Both conditions, settlement and upheaval, will be described more fully in following chapters and both are obviously influenced by soil moisture changes — a complex interaction between moisture availability and retention — plus soil character. In order to better understand the overall problem, the topics of hydrology, clay mineralogy, and soil mechanics must be viewed in the prospective of their influences on foundation design.

First, the phenomenon of natural water must be understood. Classically this study is referred to as hydrology and generally relates to the behavior of subsurface water.

CONTENTS

1

WATER BEHAVIOR IN SOILS

MOISTURE REGIMES

Subsurface water can be divided into two general classifications, the aeration zone and the saturation zone. The saturation zone is more commonly termed the "water table" or ground water, and is, of course, the deepest. The aeration zone includes the capillary fringe, intermediate belt (which may include one or more perched water zones) and, at the surface, the soil water belt, often referred to as the root zone. Refer to Figure 1-1. Simply stated, the soil water belt provides moisture for the vegetable and plant kingdom; the intermediate belt contains moisture essentially in dead storage — held by molecular forces: the perched ground water, if it occurs, develops essentially from water accumulation either above a relatively impermeable strata or within an unusually permeable lens. Perched water occurs generally after a good rain and is relatively "temporary"; the capillary fringe contains capillary water originating from the water table.

The soil belt can contain capillary water available from rains or watering; however, unless this moisture is continually restored the soil will eventually desiccate through the effects of gravity, transpiration, and/or evaporation. In so doing the capillary water is lost. This zone is also the one most critically influencing both foundation design and foundation stability as will be discussed in following sections.

As stated, the more shallow zones have the greatest influence on surface structures. Unless the water table is quite shallow, it will have little, if any, material influence on the behavior of foundations of normal residential structures. Further, the

1

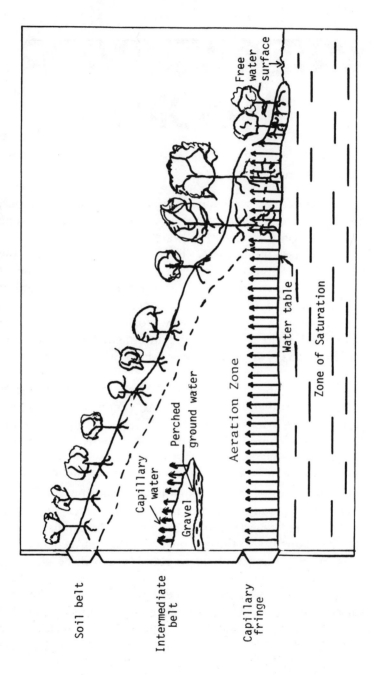

Figure 1-1. Moisture regimes.

surface of the water table (*phreatic boundary*) will not normally deflect or deform except under certain conditions in the proximity of producing well. In this instance the boundary will drawdown or recede. In other words, if the water table is deeper than about 10 ft, the boundary (as well as capillary fringe) is not likely to "dome." Should upward deflection or "doming" occur, it would more likely affect foundation than would the aforementioned draw-down. The relative thickness and depth of the various zones depends upon many factors such as soil composition, climate, geology, etc. More will be discussed on this in the following paragraphs.

SOIL MOISTURE vs WATER TABLE

Alway and McDole conclude that deep subsoil aquifers (e.g., water table) contribute little, if any, moisture to plants (and, hence, foundations). Upward movement of water below a depth of 12 in. was reportedly very slow at moisture contents approximating field capacity. The "field capacity" was defined as that residual amount of water held in the soil after excess gravitational water has drained and after the overall rate of downward water movement has decreased (zero capillarity). Soils at lower residual moisture contact will attract and flow water at a more rapid rate. Water tends to flow from "wet" to "dry" the same as heat flows from hot to cold — higher energy level to lower energy level.

Rotmistrov suggests that water does not move to the surface by capillarity from depths greater than 16 to 20 in. This statement does not limit the source of water to originate from the water table or capillarity fringe. Richards indicates that upward movement of water in silty loam can develop from depths as great as 24 in. McGee postulates that 6 in. of water can be brought to the surface annually from depths approaching 10 ft. Again the source of water is not restricted in origin.

The seeming disparity among these hydrologists is likely due to variation in experimental conditions. Nonetheless, the obvious consensus is that the water content of the surface soil tends to remain relatively stable below very shallow depths and that the availability of soil water derived from the water table ceases when the boundary lies at a depth exceeding the limit of capillary raise for the soil. In heavy soils (e.g., clays), water loss practically ceases when the water table is deeper than 4 ft even though the theoretical capillary limit might exceed this distance. In silts the capillary limit may approximate 10 ft as compared to 1-2 ft for sands. The height of capillary rise is expressed by the equation:

$$\pi \gamma_T r^2 h_c = T_{ST}^2 \ \pi \ r \cos \alpha \qquad (1\text{-}1)$$

or

$$h_c = \frac{2 \ T_{ST}}{r \ \gamma_T} \cos \alpha$$

where h_c is the capillary rise (measured in centimeters), T_{ST} is the surface tension of the liquid at temperature T (measured in grams/centimeter), γ_T is the unit weight of liquid at temperature T (measured in grams/cubic centimeter), r is the radius of the pore or capillary (measured in centimeters), and α is the angle of meniscus at the wall or angle of contact. For behavior in soils, the radius, r, is difficult, if not impossible to establish. Since the capillary rise varies inversely with effective pore or capillary radius, this value is required for mathematical calculations; however, the value of r for soils is most elusive – dependent upon such factors as void ratio, impurities, grain size and distribution, permeability, etc. Accordingly, capillary rise, particularly in clays, is generally determined by experimentation. In clays the height and rate of rise is impeded by the swell (loss of permeability) upon invasion of water. Finer soils will create a greater height of capillary rise but the rate of rise will be slower.

SOIL MOISTURE vs AERATION ZONE

Water in the upper or aeration zone is removed by one or a combination of three processes: transpiration, evaporation, and gravity.

Transpiration

Transpiration refers to the removal of soil by vegetation. A class of plants, referred to as *phreatophytes*, obtain their moisture, often more than 4 ft of water per year, principally from either the water table or capillary fringe. This group includes such seemingly diverse species as reeds, mesquite, willows, and palms. The remaining two groups, mesophytes and xerophytes, obtain their moisture from the soil water zone. These include most vegetables and shrubs along with some trees.

With all vegetation, root growth is toward soil with greater available moisture. Roots will not penetrate a dry soil to reach moisture. The absorptive area of the root is the tip where root hairs occur. The loss of soil moisture due to transpiration follows the root pattern and is generally somewhat circular about the stem or trunk. These factors are important to foundation stability as will be discussed in following chapters.

In many instances transpiration accounts for greater loss of soil moisture than does evaporation.

Gravity and Evaporation

Gravity tends to withdraw all moisture downward from the soil within the aeration zone. Evaporation tends to withdraw moisture upward from the surface soil zone. Both forces are retarded by molecular, adhesive, and cohesive attraction between water and soil as well as by the soil's ability for capillary recharge. If evaporation is prevented at the surface, water will move downward under the forces of gravity until the soil is drained or equilibrium is attained with an impermeable layer or saturated layer. In either event, given suitable time, the retained moisture within the soil would approximate the "field capacity" for the soil in question. In other words, if evaporation were prevented at the soil surface, for example, by a foundation, an "excessive" accumulation of moisture would initially result. However, given sufficient time, even this protected soil will reach a condition of moisture equilibrium somewhere between that originally noted and that of the surrounding uncovered soil. The natural tendency is for covered soil to retain a moisture level above that of the surrounding uncovered soil – except, of course, during periods of heavy inundation (rains) wherein the uncovered soil reaches a temporary state at or near saturation. In this latter instance the moisture content decreases rapidly once the source of water ceases. The loss of soil moisture from beneath a foundation caused by evaporation would tend to follow a triangular configuration with one leg being vertical and extending downward into the bearing soil with the other leg being horizontal and extending under the foundation. The relative lengths of the legs of the triangle would depend upon many factors such as the particular soil characteristics, foundation design, weather, availability of moisture, etc. In any event, the affected distances (legs of the triangle) are relatively limited as explained in preceding paragraphs. As with all cases of evaporation the greatest effects are noted closer to the surface. In an exposed soil, evaporation forces are ever present so long as the atmospheric humidity is less than 100%. The forces of gravity are effective whether the soil is covered or exposed.

PERMEABILITY vs INFILTRATION

The infiltration feature of soil is more directly related to penetration from rain or water at the surface than to subsurface vertical movement. The exception being those relatively rare instances where the ground surface is within the capillary fringe. Vertical migration or permeation of the soil by water infiltration could be approximately represented by the single-phase steady-state flow equation as postulated by Darcy:[2]

where
$$Q = \frac{-AK}{\mu}\left(\frac{\Delta P}{L} + g\,\gamma\,\sin\alpha\right) \qquad (1\text{-}2)$$

Q = rate of flow in direction L
A = cross-sectional area of flow
K = permeability
μ = fluid viscosity
$\Delta P/L$ = pressure gradient, in direction L
L = direction of flow
γ = fluid density
α = angle of dip: $\alpha > 0$ if flow L is up dip

where g = gravity constant

if α = 90 deg., sin α = 1

When flow is horizontal the gravity factor g drops out. Any convenient set of units may be used in Eq. (1-2) so long as they are consistent. Several influencing factors represented in this equation pose a difficult deterrent to mathematical calculations. For example, the coefficient of permeability, K can be determined only by experimental processes and is subject to constant variation even within the same soil. The pore sizes, water saturation, particle gradation, transportable fines, and mineral constituents all affect the effective permeability, K. In the instance of swellable clays the variation is extremely pronounced and subject to continuous change upon penetration by water; in the case of clean sand, the variation is not nearly as extreme and reasonable approximations for K are often possible. The hydraulic gradient, ΔP, and the distance over which it acts, ΔL, are also elusive values. In essence, Eq. (1-2) provides a clear understanding of factors controlling water penetration into soils but does not always permit accurate mathematical calculation. The rate of water flow does not singularly define the moisture content or capacity of the soil. The physical properties of the soil, available and residual water, each affect infiltration as well as does permeability. A soil section 3 ft thick may have a theoretical capacity for 1½ ft of water, which is certainly more water than occurs from a serious storm; hence, the moisture-holding capacity is seldom, if ever, the limiting criterion for infiltration. This is as it would appear from the foregoing paragraphs. In addition to the problems of permeability, infiltration has a inverse time-lag function (see Fig. 1-2). This figure depicts a typical, graphical representation of the relationship of infiltration and runoff with respect to time. At the onset of rain, more water infiltrates but as time progresses most of the water runs off and little adds to the infiltration.

Clays will have a greater tendency for runoff as opposed to infiltration than will sands. The degree of the slope of the land would have a comparable effect since steeper terrains would

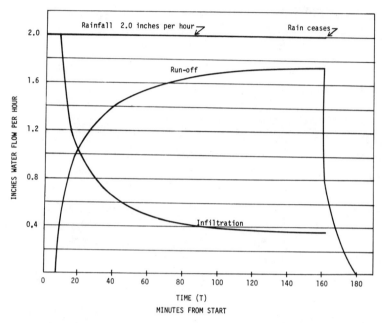

Figure 1-2. Infiltration vs. run-off after a 2 in/hr rainfall (typical).

deter infiltration. Only the water that penetrates the soil is of particular concern with respect to foundation stability; therefore, the water which fails to penetrate the soil is only briefly discussed in the following paragraphs.

Runoff

The tendency of any soil at a level above the capillary fringe is to lose moisture through the various forces of gravity, transpiration, and evaporation. Given sufficent lack of recharge water, the soil water belt will merge with, and become identical in character with, the intermediate belt. However, nature provides a method for replenishing the soil water through periodic rainfall. Given exposure to rain, all soils absorb water to some varying degree dependent upon such factors as residual moisture content, soil composition and gradation, time of exposure, etc. The excess water not retained by the soil is termed runoff (refer to Figure 1-2).

As would be expected, sands have a high absorption rate while clays have a relatively low absorption rate. A rainfall of several inches over a period of a few hours might saturate the soil water belt for sands but penetrate no more than 6 in. in a well-graded high-clay soil. A slow, soaking rain would materially increase penetration in either case.

The same analogy would hold whether the source of water is from rain or watering. Later chapters will develop the importance of maintaining soil moisture to aid in preventing or arresting foundation failures.

CONCLUSIONS

What factors have become obvious with respect to soil moisture as it influences foundation stability?

- Soil moisture definitely affects foundation stability, particularly if the soil contains swellable clays.
- The soil belt is the zone which affects or influences foundation behavior the most.
- Constant moisture is necessary for foundation safety.
- The water table, per se, has little if any influence on soil moisture.
- Vegetation removes substantial moisture from soil. Roots tend to "find" moisture.
- Moisture content under a slab foundation tends to be higher than surrounding exposed areas.
- Sources of unusual amounts of water under the foundation are accumulative and detrimental. Sources for "unusual" water could be attributed to subsurface aquifers (e.g., temporary perched ground water), surface water (poor drainage), and/or domestic water (leaks or improper watering).
- The infrequent instance of true springs is not considered herein in deference to brevity.

- Assuming adequate drainage, proper watering is absolutely necessary to maintain consistent soil moisture during dry periods — both summer and winter.

The home owner can do little to affect either the design of an existing foundation or the overall subsurface moisture profile. Speaking from a logistical standpoint, about the only control the owner has is to maintain moisture around the foundation perimeter by both watering and drainage control and to preclude the introduction of domestic water under the foundation. Adequate watering will help prevent or arrest settlement brought about by soil shrinkage resulting from the loss of moisture. From a careful study of the behavior of water in the aeration zone, it appears that the most significant contributing factor to distress is excessive water under the slab foundation (upheaval). Field data compiled from the study of over 2,000 repairs confirms this impression and offers undeniable data that a wide majority of these instances were, in turn, traceable to domestic water sources. (The numerical comparison of upheaval vs settlement failure is estimated to be in the range of 65::35 or about 2 to 1.)

BIBLIOGRAPHY

1. Meinzer, O. E. et al. *Hydrology*. New York: McGraw-Hill, 1942.
2. Pirson, S. J. *Oil Reservior Engineering*. New York: McGraw-Hill, 1958.

2
SOIL MECHANICS

The term "soil" generally describes all the loose material constituting the earth's crust, in varying proportions, and includes three basic materials: air, water, and solid particles. The solid particles have been formed by the disintegration of different rocks. The nature of this origin helps determine the soil behavior. The content of water and/or air, represented as the void ratio, further influences soil behavior. Generally, the higher the void ratio the lower the compressive or bearing strength and the higher the permeability. For purposes herein, the void ratio is defined as the ratio of combined volume of water and air to the total volume of the soil sample.

SOIL TYPES

As far as construction practices are concerned, the major classification divisions of soil are by particle size, either fine-grained or coarse-grained. The coarse-grained soils are the gravels and sands. The fine-grained soils consist of silts and clays. The unified soil classification system identifies the groups with the symbols "G" for gravels, "S" for sands, "M" for silts, "C" for inorganic clays, "O" for organic silts and clays and "Pt" for peat. In many cases, two symbols are used in combination to more precisely describe the soil, such as "GC," a gravely clay; "SM," a silty sand, and so on. The gravels and sands are further identified as "well graded" or "poorly graded" by combining the letters "W" for well graded and "P" for poorly graded (i.e., "GW," as well-graded gravel and "SP" a poorly-graded sand — "well graded" signifies a wide distribution in grain size and "poorly graded" signifies predominately one grain size.) The silts and clays are cross-classified as highly plastic, "H," or of medium to low plasticity, "L." A highly plastic, inorganic clay would be designated as "CH." An organic clay with low to medium plasticity would be labeled "OL." For a more in-depth discussion of the unifed soil classification refer to the bibliography.[1]

Sand and Gravel

The sands and gravels are the easiest soil types to distinguish. These consist of coarse particles which range in size of from 3 in. in diameter down to small grains which can be barely distinguished by the unaided eye as separate grains. Another principal parameter is whether the mixture of particles is well graded. If there is a fairly even distribution of grain sizes the soil is termed well graded. If a majority of particles are of one particular size, the soil is poorly graded. The bearing or shear strength of a sand or gravel depends solely upon the internal friction between grains. Generally speaking, bearing strengths are high and foundation failures relatively infrequent. Any settlement which might occur takes place almost immediately upon application of load and does not materially effect the stability of the foundation. Generally the shear strength increases as the grain size increases and a well-graded soil is most preferred. The sands and gravels are noncohesive indicating that there is no attraction or adhesion between individual soil particles.

Silts and Clays

Silts are finer than sands but more coarse than clays. They represent soil particles of ground rocks which have not yet changed their character into the minerals. Generally the silts are more stable than clays, with respect to construction problems. The clays represent the basic "culprit" to foundation stability. These are the finest possible particles, usually smaller than 1/10,000 of an inch in diameter. As will be mentioned in Chapter 3 the clays exhibit peculiar properties which are also deterrent to foundation stability. The silts and clays are cohesive in nature and tend to compress, deform, and creep under constant load. Silts and clays are particularly vulnerable to volumetric changes induced by moisture variation.

A simple empirical method for determining the type of soil is to disperse a quantity of soil into a container of water and note the rate of sedimentation. Sand and gravel will settle quickly — within about 30 sec. Silts will settle within 15 to 60 min. Clay particles will remain suspended for periods of several hours.

Rock

Rock is not always a superior foundation bed. The bearing qualities of rock are dependent upon such factors as the presence of bedding planes, faults, joints, weathering, cementation of constituents, etc. In many cases, the presence of "rock" as a foundation bearing results in faulty assumptions of foundation stability. So called "solid" rock, if that assumption were correct, would produce a proper foundation bearing. However, in many cases, the "solid rock" is not, in fact, solid and unexpected foundation failures result.

Fill

The preceding discussion has been limited to native or virgin soils. What happens if the construction site is not level and requires filling? The back-filled soil should be placed at approximately the same density and nature as it possessed before removal. Generally, the nature of the replaced soil is not difficult to reproduce; however, the "original" density may be more elusive. The density is an intricate balance of compaction and moisture content. With sands and gravels the problem is not difficult. With clays and silts, the problem becomes more involved. Compaction is required along with the introduction of water to the extent that maximum density is attained (see Figure 2-1).

As water is added, the first volume fills voids and helps the particles move closer together thus increasing the density.

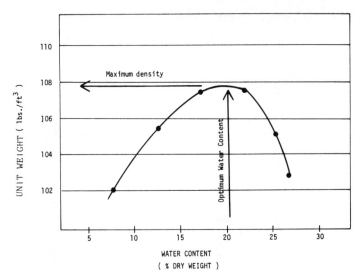

Figure 2-1. Optimum density vs. water content.

Extra water beyond the optimum displaces the heavier solids and thus reduces the density. Hence, the "optimum" water produces the maximum density at the prescribed compaction.

FOUNDATION ON SOILS

The foundation load imposed on the supporting soil depends, to a large extent, upon the design of the foundation and the character of the particular bearing soil. Figure 2-2 shows the distribution of foundation loads vs depth in bearing soils. For simplicity, the depiction relates to spread footings of a limited area. An approximate relationship would hold for strip footings, e.g., foundation beams. The width of the footing in Figure 2-2 is represented as W and the bearing load as q. The effects of the load are generally diminished at a depth of about $2W$. Immediately under the footing the load is, of course, q. At a depth of approximately ½W the load is $0.8q$. At a depth of W, the load is approximately $0.5q$. Obviously the effects of a bearing load on the underlying soil diminish rapidly with depth. This relationship will be further discussed in Chapter 4.

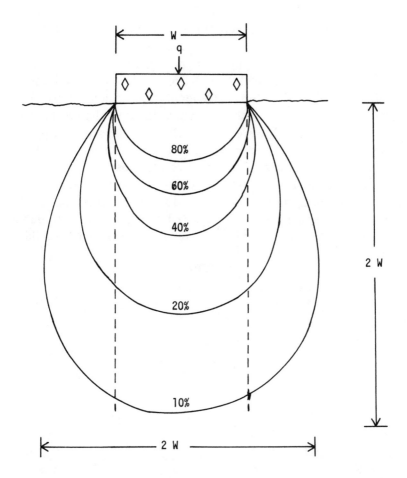

Figure 2-2. Foundation load distribution into bearing soils (Vertical).

FROST HEAVING

Frost heaving occurs when a mixture of soil and water freezes. When the soil freezes the total volume may increase by as much as 25% dependent upon the formation of ice lenses at the boundary between the frozen and unfrozen soil. This swell can create distress to foundations. As an example, the Structural Research Department of the Hydro Elective Power Commission in Toronto, Canada, produced tests which indicated that a bond between concrete and frozen soil could approach 400 lb/in.^{2}. Based on this, the uplift (heave) on a 10-in.-diameter pier in a frost zone reaching 4 ft might approximate 60,000 lb. This obviously exceeds normal foundation loads and would easily result in structural failure. Generally, the finer the soil the greater the capillary effects and, consequently, the greater the moisture content. However, relatively pure clays are not prone to frost heave because water movement is restricted, particularly near the surface where temperatures are most extreme. Silts, on the other hand, are most susceptible to frost heave. Silts are fine-grained, permeable, and do not possess the swell-propensity characteristic of clays (when subjected to available water). Obviously frost heave is not a problem in milder climates.

BIBLIOGRAPHY

1. Lambe, T.W. and Whitman, R.V. *Soil Mechanics.* New York: John Wiley & Sons, 1969.

3

CLAY MINERALOGY

INTRODUCTION

Basically the surface clay minerals are composed of various hydrated oxides of silicon, aluminum, iron, and to a lesser extent, potassium, sodium, calcium, and magnesium. Since clays are produced from weathering certain rocks, the particular origin determines the nature and properties of the clay. Chemical elements present in a clay are aligned or combined in a specific geometric pattern referred to as a structural or crystalline lattice. The particular elements present, the crystalline lattice structure and ionic substitution, account principally for the various clay classifications.

For example, chemically, the kaolinite clay mineral can be represented as $(OH)_8 \ Si_4 \ Al_4 \ O_{10}$ while the type formula for the muscovite mineral is $K_2 \ Al_2 \ Si_6 \ Al_4 \ O_{20} \ (OH)_4$.[2] This illustrates two different clay minerals based on a difference in chemical elements. (Inherently this necessitates a difference in crystalline lattice.) In another study D. G. Jacobs,[3] through X-ray diffraction studies, reported that potassium treatment of vermiculite reduced the basal spacing from 14.2Å to the illite spacing of 10.3Å. Basal spacing can be simply defined as the distance between individual or molecular layers of the clay particles. The potassium-treated vermiculite exhibited properties quite analogous with natural illite. Since the crystalline elements remained unchanged, this is an example of different clay classifications based primarily on crystalline structure. (It might be interesting to point out that the basal reduction of the vermiculite (14.2Å) to illite (10.3Å) resulted in the loss of two layers of water of hydration. In more familiar terms the treatment reduced or dehydrated the vermiculite clay).

By virtue of a loose crystalline structure, most clays exhibit the properties of moisture adsorption (and ion exchange). Among the more common clays with the tendency to swell, in decreasing order, are montmorillonite, illite, attapulgite, and kaolinite. Figure 3-1 depicts the extent of general and local abundance of high clay, expansive soils. The darker areas indicate those states suffering most seriously from expansive soil problems.

These data, provided by K.A.Godfrey, Civil Engineering, ASCE, October 1978, indicate that 8 states have extensive, highly active soils and 9 others have sufficient distribution and content to be considered serious. An additional 10 to 12 states have problems which are generally viewed as relatively limited. As a rule, the 17 "problem" states have soil containing montmorillonite which is, of course, the most expansive clay. The 10 to 12 states with so-called "limited problems" generally have soils which contain clays of lesser volatility such as illite and/or attapulgite. Represented by "dotted" coloring.

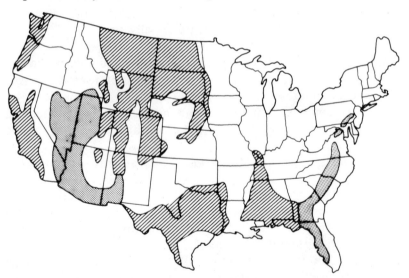

Figure 3-1. Map of U.S. shows that expansive soils are present in many of the states. Such soils are the most widespread problem in areas cross hatched. However, many locations in these areas will have no expansive soils; and in white portions of the map, some expansive soil will be found.

A specific clay may adsorb and absorb water to varying degrees — from a single layer to six or more layers — depending on its structural lattice, presence of exchange ions, temperature, environment, and so on. The moisture absorbed may be described as one of three basic forms: interstitial or pore water, surface-adsorbed water, and/or crystalline interlayer water. This moisture accounts for the differential movement (e.g., shrinking or swelling) problems encountered with soils. In order to control soil movement, each of these forms of moisture must be controlled and stabilized.

The first two forms, interstitial or pore water and surface-adsorbed water, are generally accepted as capillary moisture. Both occur within the soil mass external to individual soil grains. The interstitial or pore water is held by interfacial tension and the surface-adsorbed water by molecular attraction between the clay particle and the dipolar water molecule. Variations in this moisture are believed to account for the principal volume change potential of the soil. (It is recognized that capillary water can be transferred by most clays to interlayer water, and vice versa. However, the interlayer water is normally more strongly held and accordingly, most stable. This will be discussed at length in the following paragraph.)

In a virgin soil the moisture capacity is frequently at equilibrium even though the water content may be well below saturation.[4] Any act which disturbs this equilibrium can result in gross changes in the moisture affinity of the clay, resulting in either swelling or shrinking. Construction, excavation, and/or unusual seasonal conditions are examples of acts which can alter this equilibrium. As a rule, environmental or normal seasonal changes in soil moisture content are confined very close to the ground surface.[5] This being the case it would appear that for on-grade construction it should be sufficient to control the soil moisture only to this depth. At this point it should be emphasized that for capillary water to exist the forces of interfacial tension and/or molecular attraction must be present. Without these forces the water would coalesce and flow, under the forces of gravity, to the phreatic surface (boundary of water table). The absence of these forces, if permanent, could fix the capillary moisture capacity of the soil and aid significantly to control the soil movement. Control or elimination of retained soil moisture is the basis for chemical soil stabilization.

WATER CONTENT vs CHEMICAL STABILIZATION

Interlayer moisture is that water situated within the crystalline layers of the clay. This water provides the bulk of the residual moisture contained within the intermediate belt. The amount of this water that can be accommodated by a particular clay is dependent upon three primary factors: the crystalline spacing, elements present in the clay crystalline structure, and the presence of exchange ions. As an example, a bentonite (sodium montmorillonite), will swell approximately thirteen times its original volume when saturated in fresh water. If the same clay is added to water containing sodium chloride, the expansion is reduced to about threefold. If the bentonite clay is added to a calcium hydroxide solution the expansion is suppressed even further, to less than twofold. This reduction in swelling is produced by ion exchange within the crystalline lattice of the clay. The sorbed sodium ions (Na+) or calcium ions (Ca++) limit the space available to the water and cause the clay lattice to collapse, further decreasing the water capacity. As a rule (and as indicated by this example), the divalent ions such as Ca++ produce a greater collapse of the lattice than the monovalent ions such as Na+. An exception to the preceding rule may be found with the potassium ion (K+) and the hydrogen ion (H+). The potassium ion due to its atomic size is believed to fit almost exactly within the cavity in the oxygen layer. Consequently, the structural layers of the clay are held more closely and more firmly together. As a result the K+ becomes abnormally difficult to replace by other exchange ions. The hydrogen ion, for the most part, behaves like a divalent or trivalent ion probably through its relatively high-bonding energy.[4] It follows then that in most cases the presence of H+ interferes with the cation-exchange capacity of most clays. This has been verified by several authorities.[3,4,6] R. G. Orcutt et al. indicate that sorption of Ca++ by halloysite clay is increased by a factor of nine as the pH is increased from 2-7.[6] Though these data are limited and qualitative, they are sufficient to establish a trend. R. E. Grim indicates that this trend would be expected to continue to a pH range of 10 or higher.[4] (The clay-ion exchange at high pH, particularly with Ca++, holds significant practical importance. The pH is defined as the available H+ ion concentration. A low pH (below 7) indicates acidity. Seven is neutral and above seven is basic. Cement or lime stabilization of roadbeds represents one condition where clays are subjected to Ca++ at high pH.) It should be recognized that under any conditions the ion-exchange capacity of a clay decreases as the exchanged-ion concentration within the clay increases. Attendant with this, the moisture-sorption capacity (swelling) decreases accordingly. The foregoing discussion has referred to changes in potential volumetric expansion brought about by induced cation exchange. In nature, various degrees of exchange preexist, giving rise to widely variant soil behavior even among soils containing the same type and amount of clay. For example, soils containing Na+ substituted montmorillonite will be more volalite (expansive) than will soils containing montmorillonite with equivalent substitution of Ca++ or FE+++.

To this point the discussion has been limited to inorganic ion exchange. However, data are available which indicate that organic ion adsorption might have even more practical importance to construction problems. The exchange mechanism for organic ions is basically identical to that discussed above — the primary difference being that, in all probability, most organic sorption occurs on the surface of the clays rather than in the interlayers. Gieseking[7] reports that montmorillonite clays lost or reduced their tendency to swell in water when treated with several selected organic cations. It would appear, then, that proper utilization of the ion-exchange properties of a particular clay might be a useful tool in controlling clay moisture variations — particularly interlayer hydration or exchangeable water. This is, in fact, the basis for soil stabilization prior to the placement of foundations. At present the most common technique requires the introduction of hydrated lime, $Ca(OH)_2$, into the clay bearing soil. Chapter 7 will discuss in more depth the process of organic chemical stabilization.

BIBLIOGRAPHY

1. Terzaghi, K. *Soil Mechanics.* New York: John Wiley & Sons, 1969.

2. Bear, F. E. *Chemistry of the Soil.* 2nd ed. New York: Reinhold, 1900.

3. Jacobs, D. G. "Cessium Exchange Properties of Vermiculite." Unpublished Rept., Oak Ridge National Laboratories, Oak Ridge, Tenn.

4. Grim, R. E. *Clay Mineralogy.* New York: McGraw-Hill, 1953.

5. Meinzer, O. E. *Hydrology.* New York: McGraw-Hill, 1942.

6. Orcutt, R. G.; Kaufman, W. J.; and Klein, G. "The Movement of Radiostrontium Through Natural Porous Media." Progress Rept. No. 2, University of California to Atomic Energy Commission; November 1, 1955.

7. Gieseking, J. E. "Mechanism of Cation Exchange in the Montmorillonite — Beidilite — Nontronite Type of Clay Mineral." *Soil Science* 4 (1939): 1-14.

4
RESIDENTIAL FOUN-
DATION DESIGN

RESIDENTIAL FOUNDATIONS

A foundation is that part of a structure in direct contact with the ground and which transmits the load of the structure to the ground. This load, w, is the sum of live loads (w_L) and dead loads (w_d). The dead load is the weight of the emply structure. The live load is the weight of the building contents plus wind, snow, and earthquake forces where applicable. The magnitude of w is often assumed to be in the range of 200 to 400 lb/ft^2 for residential construction of no more than two stories. This load must be supported by the soil. That characteristic of the soil which measures its capacity to carry the load is the unconfined compressive strength, q_u. Compressible soils will normally have a q_u of less than 2,000 lb/ft^2. The actual value of q_u is determined from laboratory tests. As discussed in earlier chapters, other factors also influence foundation design such as climatic condition and bearing soil characteristics. Basically there are two different types of residential foundations — pier and beam or slab.

PIER AND BEAM

The first type foundation generally recognized for residential construction was the pier and beam. This "primitive" design involved nothing more than the use of wood "stumps" placed directly on the soil for structural support. These members, usually cedar or bois d'arc, support the sole plate as well as the interior girders. Next the design evolved to the use of a continuous concrete beam to support the perimeter, loads with, as a rule, the same wood piers serving to support the interior. This variation was precipitated largely by the advent of the brick or stone veneer which necessitated the need for a stronger, more stable base. Further changes in construction requirements, including the awareness of problems brought about by unstable soils, encouraged the addition of concrete piers to support the perimeter beam and concrete piers and pier caps, to provide better support for the interior floors (girders).

For purposes of the following discussion, pier and beam construction is that design wherein the perimiter loads are carried on a continuous beam supported by piers drilled into the ground, supposedly to a competent bearing soil or strata. The interior floors are supported by piers and pier caps that sustain the girder and joist system of a wood substructure. All foundation components are assumed to be steel-reinforced concrete. For practical purposes, there are two principal variations of the pier and beam design — the normal design wherein the crawl space is on a grade equivalent to the exterior landscape and the relatively new "low profile" where the crawl space is substantially lower than the exterior grade (see Figures 4-1 and 4-2).

The stability of the pier and beam design depends to a large extent upon the bearing capacity of the soil at the base of the piers. Assuming ideal conditions, this principle will produce a stable foundation. Difficulties with this design can occur if a stable bearing material is not accessible as a pier base. In some instances piers are belled to spread the structure load which allows the use of less competent soils to support the weight. Belled piers might provide an adequate bearing against settlement but they are somewhat vulnerable to problems of upheaval in high-clay soils. The belled area tends to anchor the pier (and attached beam). Should the upward force (upheaval) be transmitted to the beam, a cracked foundation could result. The use of void boxes under the beams has provided some limited protection against upheaval stress. However, proper moisture control is still the most successful stabilizing effect. Effective exterior grade, adequate watering, and sufficient ventilation are all forms of proper moisture control.

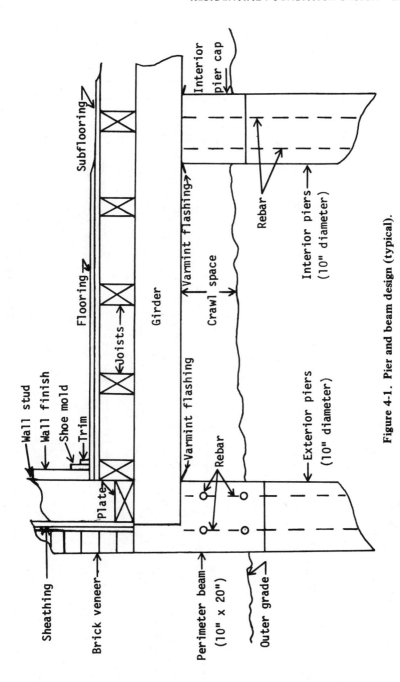

Figure 4-1. Pier and beam design (typical).

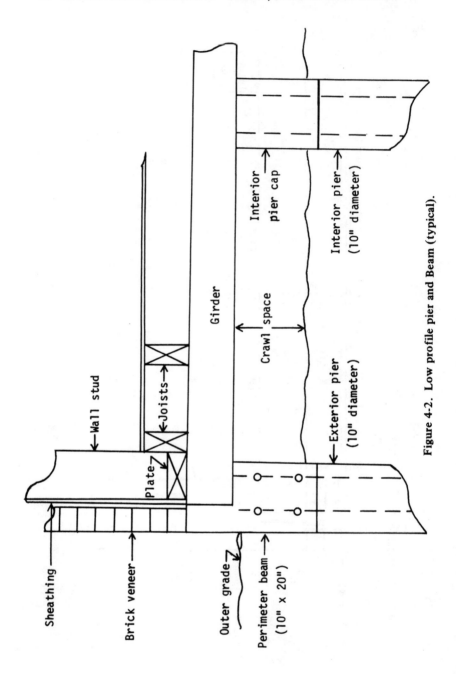

Figure 4-2. Low profile pier and Beam (typical).

The low-profile design presents inherent problems related more to excessive moisture accumulation than to actual foundation distress. The low-profile design encourages the accumulation of water under the floors which tends to result in mold, mildew, rot, termites, and if neglected, differential foundation movement. In this instance, proper ventilation and proper exterior grade will normally arrest the serious structural distress. In persistent problems of moisture accumulation, forced air blowers to provide added ventilation and/or chemical treatment of the soil to kill the mildew can be beneficial. The practice of covering the crawl space with poyethylene sheets should be avoided. Although the film reduces the humidity in the crawl space it also develops a "terrarium" beneath the sheets with subsequent threat to the foundation. In no event should the air vents be below the exterior grade. This oversight would channel surface water directly into the crawl space.

Slab

Slab construction of one variety or another probably constitutes the majority of new construction in geographic areas exposed to high-clay soils. The particular design of the slab depends upon a multitude of factors. As stated earlier, the purpose of this writing is not to delve into structural design but to present a study of foundation failures. Accordingly, the handling of slab design will be cursory. In all events the structural load is designed to be carried essentially 100% by the bearing soil. For want of a place to start the FHA (or HUD) designs pose a handy reference. Most of the FHA intelligence was apparently obtained from independent research. Fortunately much of this purchased data is available to the industry.[1] One example is the much referenced publication "Criteria for Selection and Design of Residential Slab on Ground", BRAB publication No. 33. Many other publications have been sponsored by the government but as far as analysis can determine none have substantially changed the principal design criteria postulated by the referenced BRAB publication.[1]

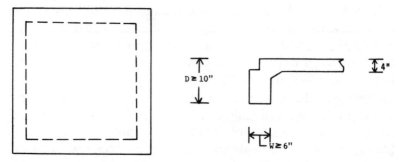

Figure 4-3. FHA Type I slab (typical).

Slab foundations come in various configurations as determined by both soil condition and the variable nature of weather (see Figures 4-3 through 4-6). As a rough "rule of thumb" soil expansion greater than about 3% is considered potentially dangerous and requires special design considerations.

Figures 4-3 through 4-6 depict typical representations of various slab designs — subject to a large extent on FHA design. Since structural design is basically beyond the scope of this article, the primary concern will be relegated to major differences in design. Basically, the resistance of a beam to differential deflection is influenced more by depth than width. In fact, other things being equal, an increase in depth by twofold will improve

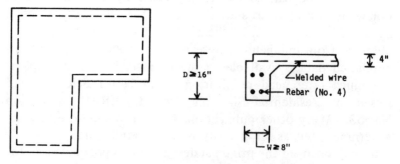

Figure 4-4. FHA Type II slab (typical).

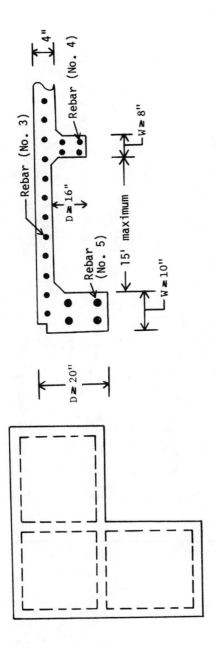

Figure 4-5. FHA Type III slab (typical).

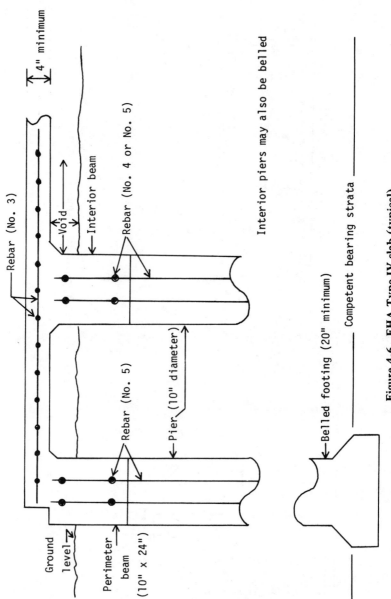

Figure 4-6. FHA Type IV slab (typical).

Table 4-1
(Simplified from BRAB Report 33)

Soil Type	Minimum Densities, PI, or q_u	Climatic Rating, C_w	Type Slab	Reinforcement
Gravel	All densities	All	I[a]	None except at openings step downs, etc., wire mesh
Gravel sands, low PI silts, and clays	Dense	All	I	Same
Same	Loose (non-compacted)	All	II[b]	Lightly reinforced, wire mesh, rebar in perimeter beam
Organic or inorganic clay and silts	PI less than 15	All	II	Same
Same	PI greater than 15	C_w greater than 45	II	Same
		C_w Less than 45	III	Rebar plus interior beams
Same	q_u/w between 2.5 and 7.5	All	III	Same
Same	q_u/w less than 2.5	All	IV	Structural slab (not supported directly on soil)

a. Maximum partition load 500 pounds per linear foot (plf).
b. Maximum partition load 500 plf unless loads are transmitted to independent footings.

the resistance to deflection by a factor of eight. Table 4-1 presents some of the major differences in soil (based on classification) and climatic conditions which combine to influence the acceptable foundation design.

Note in particular the dependence of the design upon the plasticity index, PI, and climatic condition, C_w. The PI is a dimensionless constant which bears a direct ratio to the affinity of the bearing soil for volumetric changes with respect to moisture variations. The higher the value of the PI, the greater the volatility of the soil. Generally the volumetric changes are directly related to clay content of the soil as well as differential moisture. (For example, Poor and Tucker[3] reported in their study that a slab foundation (Typical FHA "B") constructed on a soil with a PI of 42 experienced a vertical deflection of about 1 1/4 in. upon a change in soil moisture content of 4%.) The PI is determined as the difference between the liquid limit (LL) and plastic limit (PL). These terms represent part of the classical Atterberg Limit determinations. The liquid limit is determined by measuring the water and the number of blows required to close a specific width groove for a specified length in a standard liquid limit device. The plastic limit is determined by measuring the water content of the soil when threads of the soil 1/8 in. in diameter begin to crumble.[4] The letter w is the combined live and dead load for which the slab is designed and q_u is the unconfined compressive strength of the proposed bearing soil. Again referring to the BRAB report, only the top 15 ft of bearing soil influences foundation stability and the top 5 ft carries about 50% of the imposed load. Figure 4-7 shows the climatic ratings (C_w) for the United States (after U.S. Department of Commerce). In essence, as the design and environmental conditions become more severe, the slab must be strengthened. This is accomplished by increasing the size and the frequency of both the beams and reinforcing steel.

A relatively new innovation regards the use of post-tension stress to the slab. Concrete has limited tensile strength but is adequate in compression. The stress cables are intended to enhance the tensile strenth of the slab through the compressive effects of the cable tension. In theory this is fact. In practice the results do not meet expectations. Perhaps the problem lies with the fact that in field performance the compressive loads are not always perpendicular to the stress and vectors are created which destroy the benefits of the principle. The author's experiences indicate that failures with post-tension residential slabs are more likely, on a percentage basis, than with normal deformed bar slabs. As indicated above, many of these problems may be induced by field practices and, as such, are not related to the usual soil-water interaction. Figure 4-8 illustrates a typical post-tension slab.

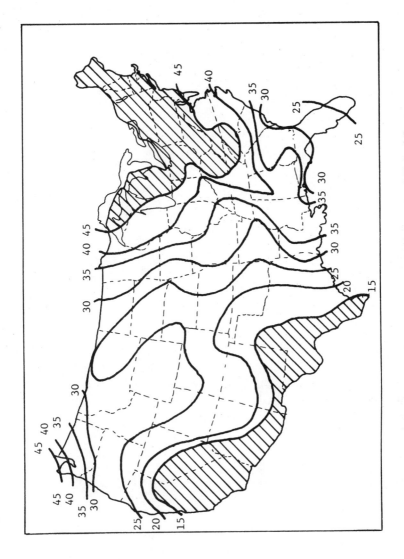

Figure 4-7. Climatic ratings (C_w) for continental United States.

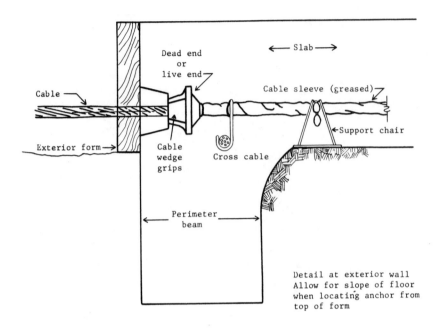

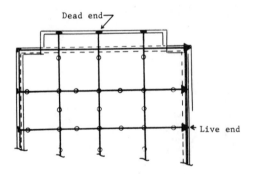

Figure 4-8. Post-tension slab (typical).

FOUNDATION LOADS ON BEARING SOILS

Generally speaking the study of foundation design, encompassing load distribution to the bearing soils, is considered beyond the scope of this book. However, since load distribution to the soil affects foundation stability, restoration of failures, and even prevention, a brief examination of foundation types will be provided. Again the comments will be limited to residential construction.

First, consider the case of the conventional spread footing. The spread footing can be part of the original construction, e.g., Type II foundations where partition loads exceed 500 plf, or incorporated into repair procedures for correcting foundation problems. One point of interest is the spread distribution of the load over the soils beneath the footing (see Figure 4-9). Assume the base of the footing at "zero." The

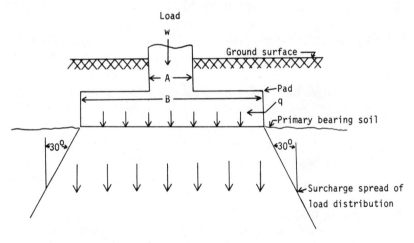

At a depth of 1.5 B the vertical stress on the
soil is still 20% of the applied load. (.2w)

Figure 4-9. Typical load distribution to bearing soil by spread footings (surcharge). (After M. J. Tomlinson.[2])

applied load is spread laterally and downward at an angle of approximately 30 deg. As the load is spread 1.73 ft vertically (down), the lateral component (out) is 1 ft. As mentioned earlier the BRAB report indicates that loads are carried 50% by soils within the top 5 ft under the foundation and 83.3% by soils within the top 10 ft. (Other authorities have indicated even shallower depths. See Figure 2-2). Hence, for all practical purposes, any stress effects below about 10 ft would be at most minimal. The lateral distance within which the soil behavior materially affects the spread footing would then be a *maximum* of (10 ft ÷ 1.73) or 5.77 ft. As a practical point, it becomes obvious that any attempt to maintain soil moisture by watering, must certainly be performed within the 5.7-ft distance. Watering beyond this distance away from the foundation would have little, if any, effect on the foundation or the foundation bearing soils, assuming of course that the exterior grade is away from the foundation as would be proper. In practice, watering to maintain moisture should be as near the foundation as practical. This is further substantiated by several authorities such as Alway, McDole, Rotmistrov, and Richards who generally agree that upward movement (evaporation) of water in silty loam does not develop from depths greater than 24 in. Restoring soil moisture at this depth under the foundation would favor watering within about 12 in.

Slab foundations distribute the loads directly to the surface (or shallow) soils. The perimeter beam carries three to four times that load borne by the interior slab. The surcharge load distribution from the perimeter beam is approximately the same as that indicated for spread footings in Figure 4-9. These points are important to remember as they influence repair techniques for correcting foundation failures.

Pier and beam load distribution is somewhat more complex. Herein, the foundation loads are carried by intermittent piers which, in turn, transmit the stresses to the soil via end bearing and skin friction (excepting clay soil where little if any design consideration is given to skin friction). Accordingly, the load distributions would deviate more from the cited examples. Nonetheless the identical theory would apply.

As suggested in foregoing paragraphs, the foundation design has some influence over the extent of or resistance to movement. The perimeter beam has the greatest structural strength and carries the greatest structural load. The only force tending to cause downward movement (settlement) is the structural load itself — with say 100–200 psf for a normal single story residence or, since about 75% of the total structured load is carried by the perimeter beam, something like 600–900 plf on the beam. Accordingly, settlement normally progresses from the perimeter to within with the greatest deflection most probable at a corner. In contrast, upheaval generally attacks the foundation's weakest point (the interior slab), and the potential disruptive force is substantially greater. The soil described by Tucker and Poor[3] in the foregoing reference can have a confined expansive force of 9,000 psf at 23% moisture content. Largely because of the gross imbalance in the applied forces and the magnitude of moisture change, upheaval normally occurs more rapidly and to a greater extent than does settlement.

BIBLIOGRAPHY

1. Federal Housing Administration. *"Criteria for Selection and Design of Residential Slab-on-Ground."* Rept. No. 33, National Academy of Sciences, 1968.

2. Tomlinson, M. J. *Foundation Design and Construction.* 2nd ed. New York: Wiley-Interscience, 1969.

3. Tucker, R. L. and Poor, A. "Field Study of Moisture Effects or Slab Movements," *Journal of the Geotechnical Engineering Division,* ASCE, April 1978.

4. Lambe, T. W. and Whitman, R. V., *Soil Mechanics.* New York: John Wiley & Sons, Inc., 1969.

5

CAUSES FOR FOUN-
DATION FAILURES

To date, much concern has been devoted to avoid foundation failures resulting from distress (or differential movement) brought about by design deficiencies or environmental extremes. Annual rainfall, temperature ranges, clay content of the supporting soil, soil composition, etc. combine to help produce the design and construction requirements for foundations. In high-clay soils, such conditions as the relative moisture content in the soil immediately prior to construction can affect the future stability or behavior of the foundation. These factors can each bear a direct influence on settlement (soil shrinkage) as well as upheaval (soil swelling). The following paragraphs will introduce an argument, supported by field results, that upheaval, not settlement, is, in fact, the most serious and prevalent deterrent to slab on-grade construction — perhaps accounting for as high as 70% of all failures.

In order to avoid any problem of semantics, it might be advised to define the terms "settlement" and "upheaval."

SETTLEMENT

Settlement is that instance in which some portion of the foundation drops below original "as-built" grade. This occurs as a result of loss of bearing — compaction of fill, erosion of supporting soil, dehydration (shrinkage of supporting soil), etc. As an example of settlement brought about by loss of soil moisture, Poor & Tucker reported a change of 1/2 in. in vertical displacement for a slab foundation over a period of about 6 months from the winter of 1971 to the summer of 1972. The

38

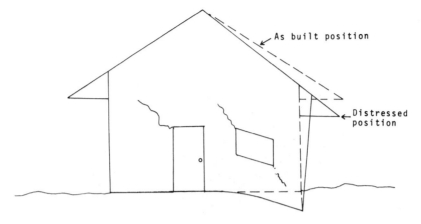

Figure 5-1. Foundation settlement (typical).

conditions of exposure (and loss of soil moisture) were more severe than one would expect for normal residential construction. The slab was completely exposed (the house had been removed) and the perimeter had not been watered. Generally, settlement originates and is more pronounced at the perimeter of the slab, and more particularly at a corner. Settlement progresses relatively slowly and can often be reversed and/or arrested by proper watering. Figure 5-1 is a typical example of foundation settlement. The dashed lines depict the as-built position and the solid lines the distressed position.

UPHEAVAL

Upheaval relates to the situation where the internal areas of the foundation raise above the as-built position. Figure 5-2 is a typical example of foundation upheaval. In high-clay soil this phenomenon results, almost without exception, from the introduction of excessive moisture under the foundation. Upheaval can develop rather rapidly depending on the quantity of available water. The author has recorded instances where 3- to 4-in. vertical deflections occurred in a slab foundation over a time interval of less than one month! Admittedly, this is un-

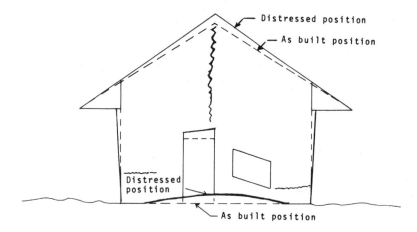

Figure 5-2. Foundation upheaval (typical).

usual but it does happen. The moisture may originate from improper drainage (faulty grade), underground water, or more frequently, from domestic sources such as leaks in supply or waste systems. Visual observations will usually reveal any grade problems.   An understanding of the particular area will help predict or evaluate the prevalence of underground water. A qualified plumbing check will verify or sometimes eliminate the possibility of any domestic leak. Watering around the perimeter can retard the noticeable progress of upheaval but will not normally reverse the problem.

The author reviewed 502 alleged foundation failures during a particular study period. The study area was primarily confined to Dallas County, Texas. Of the cases reported, 216 were no bids; that is, no problem existed which required attention other than adequate maintenance. (Most structures exhibited the result of some settlement but often this movement was not considered serious enough to warrant repairs.   In the study area, virtually all structures show the result of some movement.) The remaining 286 cases were considered to have problems sufficiently serious to warrant foundation repairs; 87 of these foundation failures were obvious, unquestioned conditions of settlement and 199 cases exhibited some indication of various degrees of upheaval. At this point, it might appear that the ratio of settlement vs upheaval would be on the basis of 87 to 199 or 30% as compared to 70%. However, as noted earlier, the field determination of settlement vs upheaval is most difficult. To the novice, the indications for both tend to appear identical. The "proof" usually relegates to the conclusive verification of water under the slab. Subsequent plumbing tests did not confirm the full 70% as will be explained in what follows.

Along these lines, the results of plumbing checks for utility leaks were compared to those instances where upheaval was suspected. The two plumbing contractors[1] involved in the study reported approximately 78.8% success in locating confirming leaks in instances where upheaval was ascribed as the cause of foundation failure. It is imperative that both supply and waste systems be carefully checked. The plumbing checks are far from certain. Positive test results will definitely confirm a leak but negative results do not necessarily prove the absence of a leak. Occasionally, after a negative test, the existence of a leak was later established when undermining the beam allowed water to run from beneath the foundation. Under normal conditions a sewer leak does not show at the ground surface.

Using a most conservative approach, the positive tests would seem to verify upheaval caused by domestic leaks in some 55% of the cases studied (70% of 78.8%).

Overall, this resolves to a statistical result depicting something like 30% failure by settlement, 55 to 65% failure by upheaval, and 5 to 15% as uncertain regarding preponderant cause.

As a point of interest, the emphasis on the plumbing check has reduced recurrent foundation failures by a factor of nearly four. Excepting previously undetected plumbing leaks, as subsequently verified by competent tests, the author has experienced foundation problems subsequent to initial repair of something slightly greater than 1%, a reduction of about 3%, from a prior experience of about 4%.

Many readers may not accept the preponderance and seriousness of upheaval. However, the available data give proof that upheaval is indeed a serious deterrent to slab on-grade construction. It is also interesting to note that in most cases the movement arrests shortly after the source for water has been eliminated. What does all this mean? Water under a foundation is truly a serious problem and must be avoided.

Perhaps the principal concern for slab on-grade foundation has been misdirected. For example, along with structural support and design, it would appear that a most important consideration would be to enact measures to control both subsurface water and surface drainage and to prevent deficiencies in the plumbing layout and design. Surface grade can be easily handled and construction in areas subject to problem subsurface water can be avoided or the condition remedied prior to construction. Simple "quality control" would probably eliminate a majority of the plumbing-induced problems. According to the plumbing contractors, whose findings and data helped to provide the basis of this chapter,[1] the principal or most common plumbing problems stem from:

1. Faulty or omitted shower pans.
2. Faulty installation of water closets (collapsed lead sleeves).
3. Faulty or ill-designed drain waste valves.
4. Ill-fitted or careless installation of connection joints.
5. Substandard sewer line materials.

In other words, added attention to the plumbing installations might reduce foundation failures in slab on-grade construction by as much as 50 to 60%. Barring this reliance, design measures could be instituted to handle any persistent deficiencies.

Bear in mind two facts. Water from leaks beneath the slab is accumulative and any water under the slab, regardless of source, tends to accumulate in the plumbing ditch. Anything that could be done to alleviate the accumulation of water in the plumbing ditch would obviously be beneficial to slab on-grade construction. There are several preventive measures which might be employed.

For example, the underground utilities could be installed in a pseudo French drain such that any effluent water would be harmlessly channeled from beneath the foundation to a convenient external disposal site. Or perhaps all plumbing trenches could be lined with an impermeable membrane.   Another possibility could be to carefully back fill around the pipes with selected material that is properly compacted. The bearing soil might be chemically treated to neutralize the affinity of the clay constituents of moisture.[2]   Perhaps better quality control at installation would minimize or eliminate leaks. Numerous other possibilities also exist.

DIAGNOSIS OF SETTLEMENT vs UPHEAVAL

Upheaval, as previously stated, is virtually always the result of some water accumulation beneath the slab. Settlement describes the condition where some area of the foundation falls below original grade. The differentiation between the two is most critical and difficult, requiring extensive experience and knowledge. Consider Figure 5-3. Did the slab foundation heave near

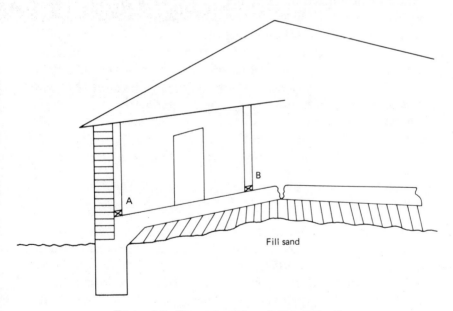

Figure 5-3. Example of foundation deflection.

the interior partition or did the perimeter settle? Consider the evidence: (1) the door frame is off 1 in. across the 30 in. header with the inside jamb being high. The grade change from point B to point A is 4.75 in. (The distance from B to A is 12 linear feet.), (2) the brick mortar joint shows a high area near the midpoint of the wall, (3) the cornice trim at both front and rear corners is out less than 1/2 in., (4) at both corners the mortar joints down adjoining walls are reasonably straight (excepting the single high area previously cited), and (5) there is no appreciable separation of brick veneer from window frames in the end walls. Analysis of the data gives rise to only one conclusion — the center has heaved. Alternately, had the perimeter settled, the greatest differential movement (4.75 in.) would show at the corners. Hence, observations 3, 4, and 5 would have been impossible. Before proper repair procedures can be established, the cause of the problem must first be diagnosed.

Often failures which are diagnosed as interior slab settlement are actually the result of a slab having been installed during a period of prolonged dry weather when the soil moisture might be abnormally low. When rains and/or domestic watering supply the perimeter areas with moisture, natural swelling occurs which raises the perimeter giving the overall appearance of interior settlement. The opposite situation (i.e., construction of a slab after extended moist weather) can occur which gives the impression of central slab upheaval when the perimeter area bearing soil returns to normal moisture content and leaves the central slab area high. In either event the vertical displacement is brought about by variation in moisture content within the expansive soil.

SLIDING

A third of type movement, sliding, sometimes occurs when a structure is erected on a slope. Herein the movement is not limited to up or down (vertical) but possesses a lateral or horizontal component. In expansive soils, cut and fill operations can be particularly precarious. The clay constituents in the cut tend to become unusually expansive even though recompacted

to (or near) their original density. This is due in large to the
breakup of cementation within the cut soil and, to some lesser
extent, the change in permeability or infiltration of the fill.

Figure 5-4 gives an example of construction on a filled slope
and Figure 5-5 shows the use of a step-beam to conform to the
contour of the land. The applied forces are basically the same in
either example and foundation failure generally results in movement down the slope. As a rule restoration of this problem is
handled as settlement.   The lower areas can be raised and
restored to a vertical position but rarely is the horizontal
component reversed.   Fortunately this type of movement is
relatively rare.

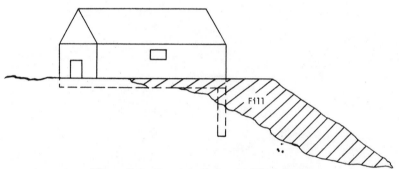

Figure 5-4. Construction on a filled slope.

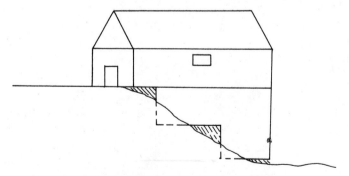

Figure 5-5. Use of step-beam to conform to contour.

BIBLIOGRAPHY

1. Brown, R. W. "A   Field Evaluation of Current Foundation Design vs Failure." *Texas Contractor*, July 6, 1976

2. Brown, R. W. et al.   "A Series on Stabilization of Soils by Pressure Grouting." *Texas Contractor*, Pt. IV, May 4, 1965

3. Brown, R. W. and Smith, C. H., Jr. "The Effects of Soil Moisture on the Behavior of Residential Foundations in Active Soils." *Texas Contractor*, May 1, 1980.

6

REPAIR
PROCEDURES

INTRODUCTION

When foundations fail, for one reason or another, prompt and competent repairs are demanded. Otherwise the problems would increase and jeopardize the value and safety of the structure. The repair procedures discussed in this chapter have been developed over the last fifteen years or so through a process of trial and error. When a technique proves successful over several thousand field evaluations, the conscientious contractor is reluctant to attempt unproven changes. In foundation repair, results far outweigh theory. As time goes on, new methods (or variations) will be introduced and, results warranting, perhaps accepted.

Before proceeding, it might be interesting to consider a few facts relative to the forces involved with leveling or raising a structure. An average, single-story brick wall with roof load exerts a downward force of approximately 10 psig. In order to lift this wall vertically, a force only slightly in excess of 10 psig is required. A 4-in. thick, open slab would require a lifting force less than 1.0 psig. (This analogy neglects the aspects of breakaway friction, mechanical binding, etc.). The shear strength of regular 5-sack concrete is probably in the neighborhood of 350 psig but might approach 1,500 to 2,000 psi. (The true value depends upon the conditions of applied stress and the investigator conducting the test.[1, 2] From these numbers it is easy to realize how structures can be raised without shearing the slab. It should be acknowledged, however, that in many cases the settlement is not uniform, and due to weight and load distribution, requires an increased and variable lifting force at different points. This is a situation requiring caution, experience, and a carefully prepared approach, without which the structure could be sheared and shattered or unevenly raised. This becomes increasingly obvious when the forces available for lifting are recognized.

The hydraulic pumps used in mud-jacking are capable of exerting pressures ranging from 100 to 500 psig — clearly in excess of that required to raise the mass of the structure. A 30-ton mechanical jack can exert approximately 8,500 psig, assuming a 3-in. head or load area. It is no wonder then that this equipment in the hands of a careless or incompetent operator can cause extensive and severe damage. On the other hand, this lifting force is sometimes required to overcome breakaway friction or mechanical binding, and, as such, must be available.

Moisture variations within expansive soils cause a preponderance of all foundation failures. The soil swells when wet (upheaval) and shrinks (settlement) when dry. This volumetric change in the bearing soil causes, in turn, differential movement in the supported foundation. According to Earl Jones and Wes Holtz in the August 1973 issue of *Civil Engineering*, 60% of all residential foundations constructed on expansive soils will experience some distress caused by differential foundation movement, and 10% will experience problems sufficient to demand repairs. Correction procedures depend both upon the nature of the problem and the type of foundation. Settlement is corrected by raising or restoring the affected area to the approximate original grade. Upheaval is corrected by raising the lowermost areas around the raised crown to a new grade sufficient to "feather" the differential.

All foundation failures are generally manifested in one or more of the following signs: interior or exterior wall cracks, ceiling cracks, sticking doors or windows, pulled roof trusses, broken windows, etc. (see Figure 6-1). A common misconception is that foundation movement occurs "instantaneously." This mistake is promoted by the fact that some of the aforementioned signs of distress do appear seemingly instantaneously.

A. Separation in exterior brick mortar.

B. Interior doors not plumb. Note the wide separation at the top right corner of the door.

Figure 6-1. (A, B, C, D) Indication of foundation distress or failure.

C. Separation of brick veneer from door jamb

D. Interior cracks in sheet rock

Figure 6-1. Continued.

However, the walls, ceilings, and other structural features are somewhat elastic in nature and ultimate failure (cracks for example) occurs when the applied stress exceeds the elasticity of the particular surface. The behavior is somewhat analgous to that of a rubber band. These indications are realized in pier and beam as well as slab foundations. In correcting any failure, one must first isolate and recognize the particular problem. For example, it is often difficult (and always critical) to differentiate between settlement and upheaval. Many of the outward signs of differential foundation movement are similar whether the actual distress is relative to settlement or upheaval. For example, refer to Figures 5-1 and 5-2: Both have cracks and both show lateral displacement of the perimeter walls. Many foundation problems have been aggravated rather than solved by errors in analysis of source of distress. However, once the cause for foundation movement has been determined, proper restitution can commence. The general delineation for the following discussions will be based on the type of foundation – either slab or pier and beam.

PIER AND BEAM FOUNDATIONS – UNDERPINNING

Pier and beam foundation movement is generally confined to settlement, though in limited cases, upheaval is observed. For practical purposes the solution for upheaval is again to treat the lowermost sections of the foundation as though settlement had occurred and raise these areas to the higher grade. Settlement is generally alleviated by mechanically raising the beam and sustaining the beam position by installing new supporting structures.

Typically, the "supports" or spread footings consist of: 1) steel-reinforced footings of sufficient size to adequately distribute the beam load and poured at a depth of be relatively independent of seasonal soil moisture variation, and 2) a steel-reinforced pier tied into the footing with steel and pored to the bottom of the foundation beam (see Figure 6-2). Design and placement of these spread footings is critical if future beam

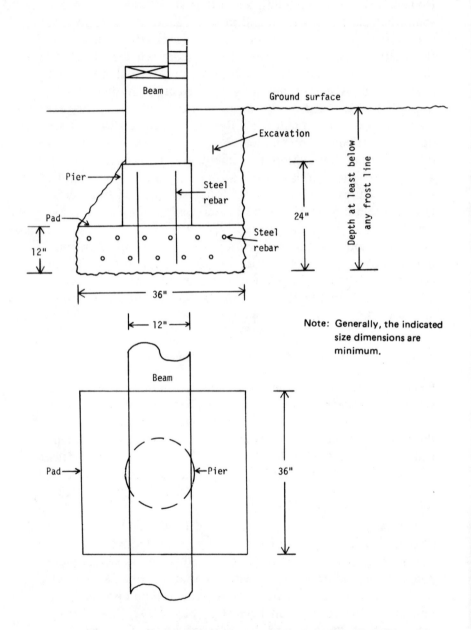

Figure 6-2. Underpinning, design of spread footing (typical).

movement is to be averted. The footing design must consider the problems of both possible future settlement and/or upheaval and should be of sufficient area to develop adequate bearing by the soil. The pier should be of sufficient diameter or size to carry the foundation load.

The foregoing paragraphs have described mechanical repairs by the use of spread footings. The principle of the footing design is to distribute the foundation load over an extended area and thus provide increased support capacity on any substandard bearing soil. The typical design represented by Figure 6-2 provides a bearing area of 9 sq ft. Effectively a load of 150 psf applied over 1 sq ft would require a soil-bearing strength of 150 psf. The same load distributed over 9 sq ft would require a soil-bearing strength of only about 17 psi.[4] This example would not apply strictly to foundations but the analogy is pertinent. The spread footings are predominately used when the overburden soil is thick and the soil-bearing strengths are low or marginal. Generally the diameter of the pier is greater than, or at least equal to, the width of the existing beam. Since the pier is essentially in compression, utilizing the principal strength of concrete, the design features are not as critical as those for the footings.

An alternate to the spread footing is the pier or piling. Generally these piers or pilings depend upon end bearing and "skin friction" (excepting in high-clay soils) for their support capacity. Piers or pilings are normally extended through the marginal soils to either rock or other competent bearing material. This obviously enhances and satisfies the support requirement. The use of piers or pilings, however, is generally restricted to instances where adequate bearing materials can be found at reasonably shallow depths. Where this is possible the installation of the piers or pilings is generally less costly than that for the spread footing. Generally speaking, for residential repairs, a competent strata depth below 8 ft precludes the use of the pier or piling technique. Upon occasion the piers or pilings are used in conjunction with the spread footing. Herein the theory is to utilize the best support features of each design with the hope of achieving a synergistic effect. In practice this goal is not normally attained. Where soil conditions dictate the spread footing, the pier provides little, if any, added benefit. However, except for cost, the integration of the deep pier as part of the spread footing has no deleterious features, provided the cross-sectional area of the footing is not diminished.

For practical purposes the residence interior is leveled by shimming on existing pier caps. This shimming will not guarantee against future reoccurrence of interior settlement since the true problem has not been solved. However, reshimming is relatively inexpensive and the rate of settlement decreases with time since the bearing soil beneath the pier is being continually compacted. True correction of the problem would normally require the installation of new piers and pier caps, at considerable expense, since the flooring and perhaps even the wall partitions would need to be removed to provide clearance for drilling.

In some instances, existing interior pier supports are unacceptable. This could relate either to concrete piers which, for one reason or another, have lost their structural competency or to deficient wood piers. In either event a proven method has been to construct new "surface" piers or pier caps. Again, the ideal solution would be the installation of "new" piers drilled into a competent bearing strata. As noted earlier this approach is prohibitively expensive for existing structures. Hence the usual approach is to provide pier caps supported essentially on the surface soil. The best compromise has been a design represented by Figure 6-3.

The pier cap depicted in Figure 6-3 involves a concrete base pad, a concrete pier cap, suitable hardwood spacers, and a tapered shim for final adjustment.

The base pad can either be poured in place or precast. The choice (and size) depends upon the anticipated load. For single-story frame construction the pad can be precast, at least 18 in. x 18 in. x 4 in. thick, with or without steel reinforcing.

In single-story brick construction and for normal two-story load conditions, the pad should be steel reinforced and at least

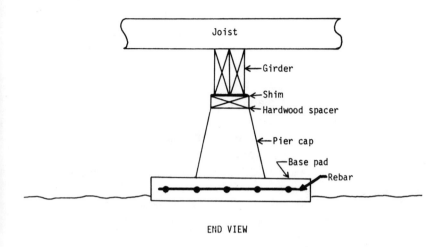

END VIEW

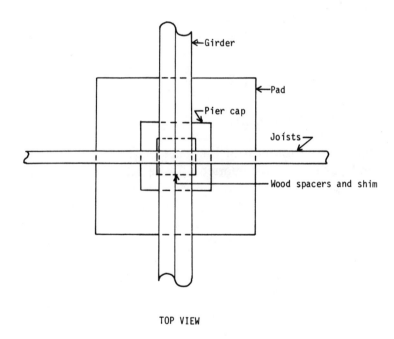

TOP VIEW

Figure 6-3. Interior pier cap (typical).

24 in. x 24 in. x 4 in. thick. For unusually heavy load areas, e.g., a multiple-story stairwell, the pad should be larger, thicker, and reinforced with more steel. In the latter case the pad is normally poured in place, due to the weight and attendant handling problems. In any event the pad is "leveled" on/or into the soil surface to produce a solid bearing. Conditions rarely warrant any attempt to place the pads materially below grade. In instances where the structure is supported on wood "piers" and/or wood stiff legs, it is also rarely justified to replace only a portion of the wood supports with the superior concrete design. Normally this would represent little, if any, benefit and be a waste of money.

The pier cap can be poured in place or precast. In most cases the limited work space demands the precast. Figure 6-3 shows one precast design. Alternatives to the precast design include cylinders, hadite blocks, square blocks, etc. Ideally, the head of the pier cap should be as wide as the girder to be supported. Either form is acceptable for selected loads. The hadite blocks, for example, are sometimes not adequate for multiple-story concentrated loads unless the voids within the blocks are filled with concrete. The precast concrete can be designed to support a load identical with the load that a pier cap poured in place can hold.

The wood spacers are hardwood and of suitable size and thickness to fill most of the space or gap between the pier cap and girder to be supported. In instances where the pier-cap head is not as wide as the supported girder, the hardwood spacers should be. If the grain of the wood in the spacer is perpendicular to the plane of the girder, the width and length can be the same. If the wood blocking is placed with the grain in line with the beam, the length should be at least twice the width. The final leveling is accomplished by the thin shim or wedge (usually a shingle). The combined thickness of the shims should normally not exceed about 7/8 in.

SLAB FOUNDATIONS — MUD-JACKING

Upheaval movement of a slab foundation presents the most serious problem to restoration. If the upward movement is pronounced, the only true cure is to break out the slab, excavate the affected base to proper grade, prepare the base, and repour the slab section. If the movement is nominal the lowermost slab areas can be raised to the grade of the heaved section. This operation is extremely critical since the grade of the entire structure is being altered. Particular care must be exercised not to aggravate the heaved area. The leveling is accomplished by what is termed mud-jacking, often in conjunction with the supplementary use of spread footings as described above. Mud-jacking is a process whereby a water and soil-cement or soil-lime-cement grout is pumped beneath the slab, under pressure, to produce a lifting force which literally floats the slab to the desired position. Figure 6-4 depicts the normal equipment

Figure 6-4. Mud-jacking equipment. The pump mixes the soil-cement with water and forces the grout through the large hose to the injection location (pre-drilled 2 inch holes).

required to mix and pump the grout. The introduction of the grout is made through small holes drilled through the concrete. This process has the attendant feature of some chemical stabilization of the soil which helps preclude future differential movement produced by soil-moisture variations (previously discussed as a preventative treatment).

Slab settlement repair is a relatively straightforward problem. Here the lower sections are merely raised to meet the original grade thus completely and truly restoring the foundation. The raising is accomplished by the mud-jack method as described above. In some instances where concentrated loads are located on an outside beam, the mud-jacking may be augmented by mechanical jacking (installation of spread footing or other underpinning). In no instances should an attempt be made to level a slab foundation by mechanical means alone. Rather than filling voids (and stabilizing the subsoil) mechanical raising creates voids which if neglected, may ultimately cause more problems than originally existed. As a rule, foundation slabs are not designed as structural bridging members and should not exist unsupported. Mechanical techniques normally provide no assistance toward correcting interior slab settlement. Mechanically raising the slab beam and back filling with grout represents a certain improvement over mechanical methods alone but also leaves much to be desired. This method loses the benefits of "pressure injection" normally associated with the true mud-jack method. In other words, the voids are not adequately filled, which thus prompts resettlement.

Figure 6-5 illustrates the effects of leveling. In this particular instance the foundation was of slab design and leveled by mud-jacking. In the "BEFORE" picture the separation under the wall partition, an indication of foundation distress, is obvious. In the "AFTER" photo the separation is closed illustrating that the movement has been reversed or corrected. In this particular example the leveling operation produced near perfect restoration. This is not always the case as discussed in various chapters within this book.

Before — the floor has dropped some 4 inches below the partition wall.

After — the floor has been raised to original grade.

Figure 6-5. Typical results from a foundation repair process.

PERMA-JACK

In 1976 a new patented process was introduced to correct failed foundations (see Figure 6-6). This technique, labeled perma-jack, utilizes a hydraulic ram to drive jointed sections of 3-in.-steel pipe to rock or suitable bearing. When the solid base is reached continued exertion of the hydraulic ram will raise the foundation. Once at grade, the pipe is pinned through the jacket bracket to hold the foundation in place. In the case of slab foundations the void beneath the slab is then filled by mud-jacking. The perma-jack process[6] is more expensive than conventional methods and in most areas will not provide superior results. Although the perma-jack process has not yet been widely accepted or used, the jobs performed using this method to correct failed foundations seem to be holding up, at least in some areas. It would appear that the technique might be useful to the foundation repair contractor in areas where the more conventional methods demonstrate problems. One basic problem with the technique is the uncertainty of alignment. One variation fills the driven pipe with concrete. This provides little, if any improvement.[7]

REVIEW OF REPAIR LONGEVITY

For a comprehensive evaluation of the relative success of foundation repairs, refer to Table 6-1. These data were collected from repair records compile essentially from Dallas, Texas.[3,8] However, similar results could be anticipated from any area, if one assumes competent repair procedures were effected. In fact, these data do include random results from Louisiana, Arkansas, Mississippi, Florida, and virtually all of Texas.

Table 6-1 emphasizes the relative occurrence of upheaval versus settlement as a cause for "redo." Also, it is obvious that the "redo" rate is quite favorable, with an overall success ratio of better than 90% (upheaval and settlement combined). These data do not delineate between slab or pier and beam foundation. However, over all, the ratio of slab repairs to pier and beam repairs was about 3 to 1.

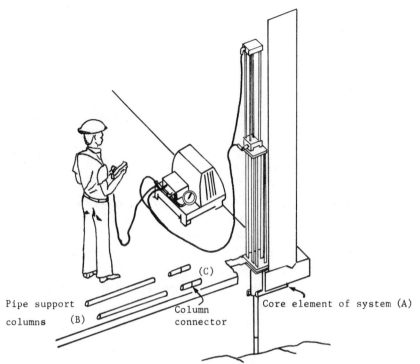

Pipe support
columns   **(B)**

(C)

Column
connector

Core element of system **(A)**

Figure 6-6. "Perma-Jack" Process.

How the "Perma-Jack" System works

An opening is made in the concrete basement floor
and enough dirt is removed so that the "Perma-Jack"
Bracket, the core element of the system (A) can be placed
under the foundation. This bracket, a strong structural
steel weldment weighing 33 pounds, is permanently installed under the foundation. Next, a pier formed from two
or more pipe support columns (B) is placed in the bracket.
These pipe support columns are aligned and tightly connected by a column connector (C). Then hydraulic force
is applied to depress this pier into the ground until it
hits bedrock.

Table 6-1.

	Success of Foundation Repairs			Cause of Failure		
Test Period	No. Jobs	Total Redos*	Upheaval	Settlement	Indeterminate
1/82–6/82	166	7 (4%)	6 (3.6%)	1 (0.4%)	—
1981	336	29 (9)	23 (7)	6 (2)	—
1980	302	36 (12)	22 (7)	14 (5)	—
Totals	804	72 (9%)	51 (6.3%)	21 (2.7%)	
12/78–12/79	243	20 (8.2)	12 (4.9)	5 (2.0)	3
12/77–12/78	291	32 (10.9)	18 (6.2)	8 (2.7)	6
Totals	534	52 (9.7%)	30 (5.6%)	13 (2.4%)	9
7/64–1/69	380	18 (4.7%)	4 (1.1%)	4 (1.1%)	10

*Excludes simple reshim of pier and beam interior floors.

In recent times when the so-called deep piers have been utilized (rather than the spread footing), the relative success does not appear to be nearly so impressive.[3]

For our purposes, the term *"redo"* refers to any rework of the foundation during the initial warranty period, usually 12 months. It might be interesting to note, however, that less than 1% of the foundations repaired developed recurrent problems after a year. At this point, a careful analysis of the data presented in Table 6-1 would be in order.

At the time the 1964–1969 data were compiled, the slab repairs were performed by mud-jacking alone — spread footings (or piers) were not utilized to supplement the raising. Further, the prevalence of upheaval was not fully recognized. The 1977–1979 data cover a span containing an inordinately wet cycle. In fact, the period of December 1977 through about May 1978 represented one of the wettest periods recorded in Texas history. The heavy precipitation gave rise to an unusually high propensity for redos, particularly upheaval. Along these lines, the overall redo rate expressed for this period is 9.7 as compared to 4.7% for the earlier period, a nearly twofold difference. Further, the relative rates of recurrent settlement vary by about the same ratio, 2.4 versus 1.1%. The significant variation involves the reported comparative incidence of recurrent upheaval, namely 5.6

versus 1.1%. This disproportionate difference is likely due to the following facts: (1) up until the 1970s, foundation repair contractors were not properly aware of the propensity of upheaval and probably mislabled some of the causes, and (2) the 1977–1979 data were accumulated during a cycle of excessive and unusual moisture, which prompted increases in upheaval. The third set of data, previously unpublished, covers the 30-month period immediately prior to July 1, 1982. Again, the influence of climatic conditions is clearly reflected. The 1980 data show the results of a record July heat wave following the extremely wet years 1978 and 1979. The year 1981 was again a near-record wet year following the unusually dry 1980. The first 6 months of 1982 represent a "return to normal" trend. These data in Table 6-1 were developed by two separate firms, the 1964–1969 data by one and the other two sets by another. Still, all factors considered, there is amazing consistency.

The incidence of upheaval as a cause for redo (as opposed to settlement) occurs at the ratio of a little over 2 to 1. At least 50% and perhaps as many as 75% of the redo cases were brought about by the refusal or neglect by the owner to properly care for or maintain the property. Along these lines refer to Chapter 9. Under normal climatic conditions the anticipated chance for any type of redo should be about 4 out of 100 jobs performed (96% success). Since upheaval is caused by excessive accumulation of water beneath the foundation, any measures that prevent this will minimize the chance for redo. For that matter, the same precautions will help avoid the initial problem as well.

CONCLUSIONS

Based on the foregoing, it becomes apparent that: (1) foundation repairs tend to principally protect against recurrent settlement, (2) recurrent upheaval is a concern with slab foundations, unless proper maintenance procedures are instituted concurrent with or prior to proper foundation repairs, (3) spread footings appear superior to "deep piers" as a safeguard against resettlement,[3] (4) "deep piers" may, in some instances, be conducive to upheaval,[3] (5) upheaval accounts for more foundation failures than settlement, (6) moisture changes which influence the foundation occur within relatively shallow depths, (7) the effects of upheaval distress occur more rapidly and to a greater potential extent than does settlement, (8) the cause of foundation problems must be diagnosed and eliminated prior to repairs if recurrent distress is to be avoided, and (9) weather influences foundation behavior.

BIBLIOGRAPHY

1. Gutschick, K. A. "Some Market Development, Success." Presentation at National Lime Association Convention, Dorodo Beach, Puerto Rico, April 26, 1967.

2. Gutschick, K. A. "Lime Stabilize Poor Soils." Concrete Construction, May 1967.

3. Brown, R. W. and Smith, C. H., Jr. "The Effects of Soil Moisture on the Behavior of Residential Foundations in Active Soil." *Texas Contractor,* May 1, 1980.

4. Brown, R. W. "A Series of Stabilization of Soils by Pressure Grouting." *Texas Contractor,* Pt. 1A, Jan. 19, 1965; Pt. 1b, Feb. 2, 1965; Pt. 2B, March 16, 1965.

5. Brown, R. W. "Concrete Foundation Failures." Concrete Construction, March 1968.

6. Langenbach, G. F. "Apparatus for a Method of Shoring a Foundation." Patent No. 3,902,326, Sept. 2, 1975.

7. Petry, Thomas, Foundation Seminar. UTA, May 20, 1983.

8. Brown, R. W. "A Field Evaluation of Current Foundation Design vs. Failure," *Texas Contractor,* July 6, 1976.

7
SUPPLEMENTAL
REPAIR
PROCEDURES

This chapter deals with special techniques available to handle the relatively unusual foundation problems. The principal areas covered are deep grouting with several variations, French drains, and forced water techniques designed to preswell the clay.

DEEP GROUTING

For purposes of the following discussion, grouting operations include any activity whereby the success of the operation depends upon the ability of an injected material to penetrate and permeate a relatively deep soil bed. Once in place, the grout must either set into a strong, competent mass or intersperse with and into the soil to create a strong, competent mass. Specifically these operations include: sub-grade waterproofing, soil stabilization for control of bearing strength or sloughing, and certain back-filling operations. Often this technique is also coupled with mechanical leveling or hydraulic stabilization to correct foundation problems originating deep below the soil surface. Upon occasion grouting is used to "freeze in" foundation piers or stabilize fill beneath foundations.

For optimum penetration, the consistency (viscosity) of the invading grout should be controlled. For maximum penetration, the consistency should be a minimum. In other cases it may be desirable to restrict penetration by increasing the consistency. In this latter case it is normally sufficient to merely decrease the water-to-solids ratio. In the former problem the solution is not so simple. While grout consistency can be decreased by increasing the water-to-solids ratio, this approach is highly restrictive, i.e., the increase in water content will drastically reduce compressive strength of the set grout and the additional water creates severe handling problems brought about by separation between solids and liquid during transportation (pumping). It is apparent that, as far as reduced consistency is concerned, another control method must be selected. One solution would be the development and use of special chemical additives.

What Causes the Slurry Consistency?

The initial consistency (thickness) of a cement or soil-cement slurry is produced by the hydrating cement particles. These hydration particles, particularly the calcium hydrosilicates, have a positive surface charge and are hydrophilic (attract water). A negatively charged chemical which would absorb on the surface of the cement particles would inhibit the attraction and exposure of these particles to the water. This would deter the formation of the hydration products normally associated with the gelation of the cement or grout slurry and produce a marked reduction in the overall consistency of the slurry. Most cement-dispersing agents exhibit secondary dispersing qualities through air entraining tendencies. (As a matter of fact, it is believed that with many of the recent, more effective, dispersants the principal dispersing qualities are realized from their air entrainment and secondly from the surface phenomenon.) A certain amount of air entrainment should also aid, to some degree, in controlling the consistency problems. Two of the more common dispersants are discussed in following paragraphs.

Evaluation of Chemical Additives

One of the more common chemical additives is calcium ligno-sulfonate (CLS), a member of the lignosulfonate family. This chemical has been used as a dispersant (or fluidifier) in the construction industry for many years. This material is distributed under a variety of trade names — sometimes in essentially pure form and sometimes blended with other additives such as aluminum powder or calcium chloride. Another product selected for evaluation was a sulfonated naphthalene, coded SFR-3 (slurry-friction reducer). Both products were evaluated in two grouting compositions:

1. A neat cement slurry containing approximately 5 gal of water per sack of cement.
2. A soil-cement grout containing 4 sacks of cement, 1,800 lb of siliceous soil, and 72 gal of water per cubic yard of a dry mix.

(The same Type I cement was used throughout and the percentages of the additives were based on the weight of the cement.) In these tests the consistencies, represented as n and k, were measured on a Fann Model VG 35 viscometer which measures viscosity (or resistance of a fluid to flow). Verification of the dependence of slurry consistency on the values of n and k appears in the following paragraphs. The slurry exhibiting the best flow characteristics occurs as n approaches 1.0 and k approaches 2.0×10^{-5}. The results of these tests are depicted in Table 7-1.

Higher percentages of CLS were avoided because of the known tendency of this material to severely increase the time required for the cement to attain initial set. In all cases the presence of the soil interfered with the activity of the chemical additive. This was caused by the tendency of particular soil constituents to compete with the cement for adsorption of the chemical. This situation would not exist with all soils.

These data show that the CLS material does produce an improvement over the untreated grout both with neat cement and soil cement. However, the SFR-3 shows a much greater relative improvement. As a matter of fact, the neat slurry treated with 1.5 to 2.0% of SFR-3 exhibits flow properties approximating that of water. Hence a grout slurry containing the proper percentage of SFR-3 could be expected to penetrate and permeate a soil area with somewhat the same efficiency as water, provided, of course, that the pore channels within the soil are

Table 7-1. Evaluation of Flow Characteristics vs Slurry Composition[a]

	n	k
1). Neat Cement Grout		
no additive	0.234	0.345
0.3% CLS	0.37	0.104
0.7% CLS	0.53	0.019
1.0% CLS	0.58	0.015
0.3% SFR-3	0.314	0.21
0.7% SFR-3	0.46	0.071
1.0% SFR-3	0.521	0.0213
1.5% SFR-3	0.948	0.00073
2.0% SFR-3	1.11	0.00018
2). Soil-Cement Grout		
no additive	0.338	0.597
0.3% CLS	0.497	0.161
0.7% CLS	0.527	0.111
0.3% SFR-3	0.491	0.155
0.7% SFR-3	0.51	0.117
1.0% SFR-3	0.636	0.05
1.5% SFR-3	0.76	0.02

[a]Data Compiled by the Western Co., Research Division, Dallas, Tex.

sufficiently large to permit passage of the grout solids. This often does not pose much of a problem since the cement and soil particles used for grouting operations are extremely small.

(The same relative improvement would be anticipated for any other grout or concrete composition containing cement since the chemical activity of the additive is directed to the cement constituent).

Before arriving at any real conclusions, however, two other aspects should be evaluated – strength development and ease of handling. The above laboratory tests, plus subsequent field tests established the fact that the use of either additive (CLS or SFR-3) did not create any particular handling problems.

Compressive strengths were determined using the grout composition specified in Table 7-1. These tests were run on 2 in. cubes according to accepted API standards, cured at 72°F for 1, 3, and 7 days. These data are depicted in Table 7-2.

Table 7-2. Evaluation of Strength Properties[a]

Slurry Composition	Compressive Strength		
1). Neat Cement Grout			
	1 day	3 days	7 days
no additive	1225	3200	5925
0.3% CLS	550	2435	3350
0.7% CLS	125	1170	3100
1.0% CLS	0	–	2500
0.3% SFR-3	875	2450	4600
0.7% SFR-3	1025	3350	5100
1.0% SFR-3	950	3125	4875
1.5% SFR-3	675	2390	3950
2.0% SFR-3	430	–	7300
2). Soil-Cement Grout			
no additive	45	140	230
0.3% CLS	45	110	200
0.7% CLS	30	110	215
1.0% CLS	0	–	180
0.3% SFR-3	45	142	213
0.7% SFR-3	40	120	218
1.0% SFR-3	40	125	203
1.5% SFR-3	40	127	200
2.0% SFR-3	40	–	222

[a] Data Compiled by the Western Co., Research Division, Dallas, Tex.

Note particularly the 7-day strengths for 2% SFR-3 neat grout, 1% CLS neat grout, and the untreated neat cement. The SFR-3 slurry developed approximately three times the strength of the CLS mix and was 23% stronger than the mix with no additive. These tests were repeated with identical results.

Note also that with 1% CLS neither test mix developed measurable strength. Even if prior n and k evaluation had indicated beneficial results at higher CLS percentages, this lack of strength development would obviate field usage. (Reference to Table 7-1 shows that the decrease in k became negligible as the percent CLS was increased from 0.7 to 1.0.)

These data further verify that SFR-3 is an acceptable product, clearly superior to CLS, and will facilitate grout placement in any grouting operation — particularly in those instances where injection is limited by low-soil permeability.

Mechanics of Grouting Applications

The specific mechanics required for a grouting operation depend entirely upon the type of operation involved. For certain grouting it is necessary to form a continuous consolidated boundary or area of sufficient strength, often referred to as a grout curtain, to permit excavation or provide adequate bearing strength. The hole pattern used to introduce the grout will depend upon the particular problem and each project should be carefully designed and engineered.

In general, however, the fewer the holes and the wider the spacing the more economical the operation. For this to be practical the grout must flow efficiently; thus, once again, pointing up the need for a consistency additive. For all practical purposes the identical situation exists for subgrade waterproofing, slab leveling, etc. It should also be pointed out that for slab leveling, back filling around conduits, filling voids beneath slabs, etc., the use of SFR-3 provides the attendant feature of reducing shrinkage in set product. This reduction of shrinkage is facilitated both by the reduced water requirement and the entrained air. If desired, other products are available in the sulfonated naphthalene family which provide greater air entrainment. The SFR-3 entrains about 5% air or less based on the total slurry volume. Other varieties of sulfonated naphthalenes will entrain up to about 15% air. For some operations the higher air content could be a decided advantage.

Importance of SFR-3 to Construction in General

Quite incidentally, the foregoing tests brought to light two other important aspects inherent to the SFR-3 material, i.e., the material's ability to 1) reduce the line pressures during concrete placement by pumping and to 2) reduce water requirements for construction concrete (increase concrete strength at constant slump). During the period that the foregoing tests were being completed a problem was presented wherein it was required to reduce the line pressure on a concrete placement pump. Immediately the thought occurred to test the SFR-3 for this application. (Line pressure (friction) is known to be proportional to fluid viscosity (slurry consistency). Accordingly, any decrease in fluid viscosity would result in decreased required line pressure.)

For this test, redi-mix concrete containing 50% 1-in. minus aggregate, 50% sand, and 5 ½ sacks of cement was circulated through the concrete pump. Initially the slump of the concrete was 2 ½ in. Upon adding 4 lb SFR-3 per cubic yard of concrete, the slump was increased to an indeterminate value − at least 10 in. − and the line pressure was decreased by about 60-70%. (In this test the concrete was overtreated. It is estimated that less than half as much additive should have been used. However, conditions were such that an additional test was impractical. The accurate pressure decrease was difficult to determine due to continual pressure fluctuations or surges, even at constant pump rates. Cylinders were collected from both the treated and untreated mix. Compressive strength texts taken at seven days established that the concrete strength remained essentially unchanged.

From these data it would appear that SFR-3 could be utilized to decrease water requirements of concrete at a constant slump. Some years back a series of tests were conducted using a concrete mix containing 7 gal of water per sack of cement for a specific slump and a seven-day-compressive strength of 3,000 psi. With this additive only 5.0 gal of water were used and the concrete developed a seven-day strength of 5,000 psi, yet maintained the specified slump. This feature could have tremendous impact on the construction industry.

Classification of Fluids – Mathematical Analysis:

Table 7-1 presented the flow characteristic values n and k inherent to two grout systems; however, nothing was mentioned regarding the theory behind the determinations. For the more serious student of flow behavior, rheology, the following development should be of interest and benefit.

Fluids can be classed into one of two groups – Newtonian or non-Newtonian. Newtonian fluids (e.g., water, antifreeze, salt water, etc.) produce a straight line through the origin when a shear rate-shear stress diagram is plotted. Stated another way, this means that as the shear rate is increased the shear stress increases proportionately. This relationship can be described by a single constant – viscosity. Mathematically, the shear rate vs the shear stress of a Newtonian fluid may be described by the following equation:

$$T = \mu\left(\frac{dv}{dy}\right) \qquad (7\text{-}1)$$

where T is the shear stress (lb/ft^2); $\frac{dv}{dy}$ is the shear rate (sec^{-1}); and μ is the viscosity (lb-sec/ft^2).

The shear stress vs the shear rate of a non-Newtonian fluid (e.g., cement, soil or soil-cement slurries) is not represented by the preceding simple equation. In order to describe a non-Newtonian fluid it is necessary to use at least two constants. One of the simplest methods which represents many fluids over a fairly large range of shear rate is the so-called power-law relationship.

$$T = k\left(\frac{(dv)}{(dy)}\right)^n \qquad (7\text{-}2)$$

where k is the consistency index (lb-sec/ft^2); and n is the flow-behavior index.

The values of k and n for a given fluid are readily determined from commercial concentric cylinder viscometers such as the Fann VG 35 instrument. In Figure 7-1, the stress on the fluid (dial readings on the instrument) is noted at various fixed rates of shear (gear setting) and the results are plotted. The slope of the resulting curve is n and the x-axis (abscissa) intercept is k. Often only two readings are taken — the readings at the 300 rpm shear rate and the reading at the 600 rpm setting. This curve, extrapolated to zero rate of shear, will provide the approximate values for k and n as indicated by the dashed line in Figure 7-1.

Also, equations are available which permit mathematical calculation of n and k from the 300 and 600 rpm readings.[1]

From the relationship introduced by Eq. (7-2) and Figure 7-1, it is apparent that the fluid characteristics depend to a large extent upon the particular shear conditions at which they are determined. For this reason it is most desirable to determine n and k at or near the conditions of stress anticipated for the particular field application. The commercial viscometers generally operate at relatively low and fixed rates of shear. This indicates that for most field operations another method for experimentally determining k and n should be available. A capillary viscometer such as a Burrell rheometer or pipe-flow apparatus can suffice. The pipe-flow apparatus represents the simplest capillary viscometer and is, as expected, the most commonly used. This equipment permits shear-stress measurements at practically unlimited rates of shear. For capillary viscometers the equation analogous to Eq. (7-2) is:

$$T_w = k'\left(\frac{(8v)}{(d)}\right)^{n'} \tag{7-3}$$

where
$\quad T_w \quad$ is the shear stress at wall (lb/ft^2);
$\quad v \quad$ is the average velocity in capillary (ft/sec.); and
$\quad d \quad$ is the diameter of capillary (ft).
$\quad k' =$ consistency index, $-sec/ft^2$
$\quad n' =$ flow index.

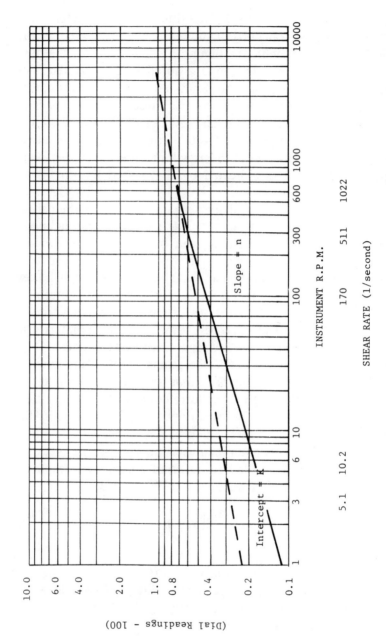

Figure 7-1. Shear stress vs shear rate as determined from concentric cylinder viscosimeter.

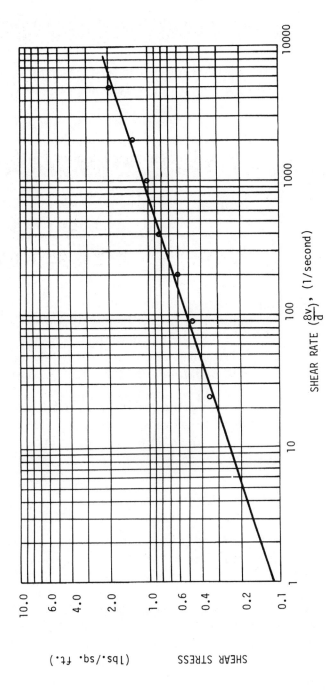

Figure 7-2. Shear stress vs shear rate from pipe flow viscosimeter.

In this method the shear rate $(8v/d)$ is a function of fluid velocity in the capillary and the shear stress $(dP/\Delta L)$ is determined by measuring the pressure differential in the capillary between two points L distance apart. These data are then plotted as previously shown by Figure 7-1 (see Figure 7-2).

k' and k as well as n and n' are closely related, the primary difference being the method used to obtain the values. For power-law fluids the correlation between concentric cylinders and capillary viscometers may be shown by the following:

$$n = n'$$

$$k = k'\left(\frac{4n'}{3n+1}\right)^{n'}$$

(7-4)

It should be noted that for any Newtonian fluid, $n = 1$ and k is the viscosity. For water, $n = 1$ and k is equal to approximately 2.0×10^{-5} –sec/ft.^{2} The greater deviation of n or n' from unity, the more non-Newtonian the behavior of the fluid.

It is recognized that the above information is brief and somewhat specialized. The purpose herein is to provide only enough insight to enable a general comprehension of non-Newtonian behavior and not to develop rheologists. Pressure grouting can be performed without this knowledge but a basic understanding of the preceding information will permit greater flexibility in field techniques.

SUBGRADE WATERPROOFING AND SOIL CONSOLIDATION

Several applications for deep grouting were mentioned earlier in the chapter but were not discussed in any detail. Since the processes of deep grouting are more common to repair procedures for commercial foundation, the following presentation will be brief and limited to those applications more likely to involve residential foundations. One example of deep grouting would be the waterproofing of subgrade basements and

and foundation walls. Another example would be the deep grouting to fill voids and increase the bearing qualities of deep fills — particularly organic fills.

Waterproofing subgrade walls is a rather routine, straightforward operation. The real difficulty lies in the fact that the stoppage of water intrusion at one point often results in the reappearance of water at unpredictable new locations. Technically, subgrade waterproofing is not materially different from regular mud-jacking, the differences being that subgrade waterproofing: 1) rather than pumping through a floor slab the grout is injected through the walls; 2) working or injection pressures are higher; and 3) the grout composition is often modified, principally by increasing the cement content. Since each situation is different the appropriate repair technique must be signed for the specific problem.

The results obtained from this process are generally satisfactory. Upon occasion, as noted above, the intrusion of water may reappear at a new location or the water flow may be reduced to an intolerable seepage. In either of these instances the grouting operation is normally repeated. From cases within the personal knowledge of the author, the initial procedure was successful in over 80% of the cases. Less than 5% of the cases required a third application and none required a fourth. Success varies directly with technique, nature of the water problem, and, upon occasion, particular job or site conditions.

Consolidation of deep fill by pressure grouting represents a more involved operation. The grout must be placed to fill voids as well as permeate the soil matrix to a sufficient extent to develop the desired bearing strength of the fill segment. The quality and nature of the fill material dictates the grout composition. Often a cement grout, with or without chemical additives, is acceptable. In any event, the selected grout is mixed and pumped into the selected site. Generally the placement commences at the lowermost elevation and proceeds upward to the surface or to some predetermined level. Consistent with this, the grout pipe is placed at the bottom of the zone to be grouted and progressively raised in lifts to accommodate the upper levels. Pumping normally continues at each level until either some selected resistance pressure is encountered or undesired movement is detected at the surface. Figure 7-3 illustrates a typical pressure-grouting operation.

The represented "void areas" are not meant to imply that either voids or incompetent sections follow any defined pattern. The grout placed alternately may develop into a "bulb" configuration around the grout pipe with little or no penetration into the soil. This might well be satisfactory, provided the placement pressure is sufficiently high to cause either mechanical consolidation of the surrounding fill or if the grout permeates and penetrates the matrix soil to the extent that sufficient "cementitious" consolidation occurs. The so called "fringe"

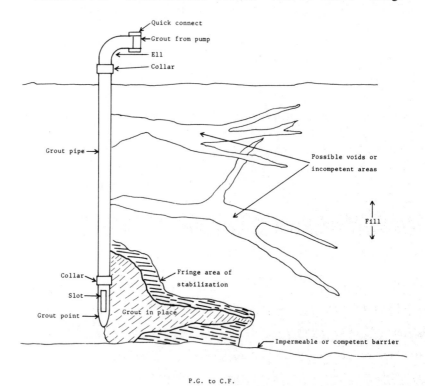

P.G. to C.F.

Figure 7-3. Pressure grouting to consolidate fill.

area of stabilization as depicted in Figure 7-3 may or may not occur, dependent upon the particular properties of the specific grout. Generally, the aqueous phase of the grout is responsible for this benefit which often results in a significant improvement to the competency of the material.

Upon occasion stage grouting is required to provide the desired results. Here the grouting operation is merely repeated until acceptable consolidation is achieved. Generally, no more that two to three stages are required within a given area to produce the desired results.

FRENCH DRAINS

Upon occasion, subsurface aquifers permit the migration of water beneath the foundation. When the foundation is supported by a volatile (high-clay) soil, this intrusion of unwanted water must be stopped. The installation of a French drain to intercept and divert the water is a useful approach. As shown in Figure 7-4 the drain consists of a suitable ditch, cut to some depth below the level of the intruding water. The lowermost part of the ditch is filled with gravel surrounding a perforated pipe. The top of the gravel is continued to at least above the water level and often to the surface. Provisions are incorporated to remove the water from the drain either by a gravity pipe drain or a suitable pump system. Simply stated the French drain creates a more permeable route for flow and carries the water to a safe disposal point. If the slope of the terrain is not sufficient to afford adequate drainage, the use of a catch basin-sump pump system is required.

The subsurface water normally handled by the French drain is the aforementioned perched ground water but upon occasion the lateral flow from "wet weather" springs or shallow aquifers is also accommodated.

Where the conditions warrant, the design of the drain can be modified to also drain excessive surface water. This is readily accomplished by either adding surface drains connected into the French drain system or carrying the gravel to the surface level.

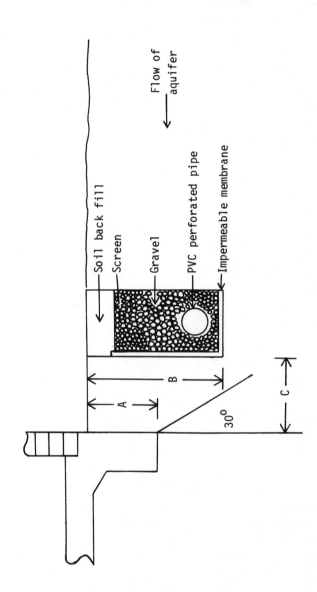

Figure 7-4. French drain (typical). Generally, C is equal to or greater than A, and B is greater than (A + 2 feet). The drain should be outside the soil bearing surcharge area.

A French drain is of little or no use in relieving water problems resulting from a spring within the confines of the foundation since it is almost impossible to locate and tap a spring beneath a foundation. A proper drain intercepts and disposes of the water before it invades the foundation.

When a foundation problem exists before the installation of a French drain, foundation repairs are also often required. In this event the repairs should be delayed to give ample time for the French drain to develop a condition of moisture equilibrium under the foundation. Otherwise, recurrent distress can be anticipated.

EXTREME MOISTURE CONTROL (INDUCED)

As stated in preceding chapters, variations in moisture content represent the principal cause of foundation distress. Should the moisture content remain stable, the clay constituent does not produce the volume change. Presumably, the easiest method for maintaining a constant moisture level would be to keep the soil either completely dehydrated or totally saturated. Attempts have been made in the past to utilize this theory to achieve stability of residential foundations — particularly with slabs. The general approach has been to flood the foundation (saturation). As far as available data indicate, the attempts have not been particularly successful. The following paragraphs have been included to allow the reader the opportunity to evaluate the potential of the extreme moisture controls.

Saturation

The intended objective is to "preswell" the clay to the maximum volume by providing sufficient water to obtain total saturation. The soil within the confines of a slab foundation can be readily saturated by water injection; however, the natural tendency persists for the moisture to migrate laterally and thereby escape at the periphery. This natural loss of water can be retarded by continued watering at the perimeter, but the loss cannot be completely stopped. Hence, the moisture content beneath the slab varies, accompanied by corresponding volumetric changes within the soil. Couple this with the inherent inconsistency of the soil-swell potential over the foundation area and the fallacy of moisture control at or near saturation becomes obvious. One method to help overcome the peripheral loss of water has been to install a vertical water barrier around the perimeter as indicated by Figure 7-5. Herein, the membrane, polyethylene as a rule, is installed in a slit trench, parallel to the beam, and at sufficient depth to prevent the exodus of the water. Generally this depth should be at least 2 ft below the bottom of the beam. Even if the exceptional moisture content could be maintained, the variable nature of the soil would most often prevent the process from being adequately successful or satisfactory. The water barrier might also have some application to prevent the intrusion of water from the outer perimeter into the foundation bearing soil. However, this benefit would be limited and could be handled by other techniques. In this connection, a vertical water barrier approximately 5–6 ft deep and utilizing a lean concrete mix, as opposed to the polyethylene membrane, may be used to prevent penetration of moisture robbing tree roots under the slab.[3] This procedure would also serve the secondary purpose described above for controlling the flow of soil moisture.

Dehydration

Attempts to attain a prolonged condition of complete dehydration are equally elusive. Heavy rains, excessive watering, bad drainage, or domestic leaks would each preclude the possibility of maintaining the dehydrated state.

The practical solution is to accept the "existing" state of moisture, somewhere between 0% (dehydration) and 100% (saturation) and make all reasonable efforts to maintain these levels. Since the moisture is admittedly subject to variation,

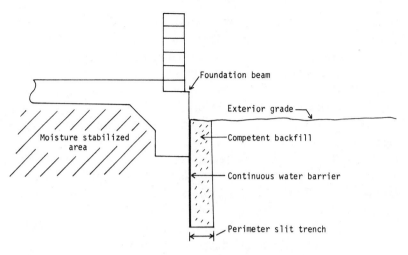

Figure 7-5. Water barrier.

some foundation movement is likely. However, the distress can hopefully be contained within tolerable limits. This is the approach accepted and suggested in this book.

CHEMICAL STABILIZATION

Chapter 3 introduced and briefly described the principle of soil stabilization by treatment with selected chemical agents, which generally relegates to the use of calcium hydroxide (lime). While chemical stablilzation is effective both as a preventative treatment in new construction and as a remedial treatment to arrest foundation distress, the following discussion will be focused toward the latter.

From a broad view, certain chemical qualities tend to help stabilize expansive clays. Among these are: high pH, high [OH] substitution, high molecular size, polarity, high valence (cationic), low ionic radii, and highly polar vehicle. Examples of organic chemicals which possess a combination of these features are: polyacrylamides, polyvinylalcohols, polyglycolethers, polyamines, polyquarternaryamines, pyridine, collidine, and certain salts of each.

Since none of the above organic chemicals possess all the desired qualities, they are often blended with other additives to enhance their performance. For example, the desired pH can be attained by addition of hydrochloric acid (HCl) or acetic acid (C_2H_3 OOH); the polar vehicle is generally satisfied by dilution with water; high molecular size can be accomplished by polymerization; surface active agents can be utilized to improve penetration of the chemical through the soil; and so on.

Generally, the organic chemicals can be formulated to be far superior to lime with respect to clay stabilization. The organic products can be rendered soluble in water for easy introduction into the soil. Chemical characteristics can be more finitely controlled, and the stabilization process is more nearly permanent. About the only advantage lime has over specific organics is, at present, lower treatment costs, more widespread usage (general knowledge), and greater availability.

The point was made in foregoing discussions that foundation repair generally is intended to raise lowermost areas of a distressed structure to produce a more nearly level appearance. The repair could be expected to be permanent only if procedures were implemented to control soil moisture variations. This is true because nothing within the repair process will alter the existing conditions inherent to an expansive soil. Alternately, chemical stabilization tends to alter the soil behavior by eliminating or controlling the expansive tendencies of the clay constituents when subjected to soil moisture variations. If this reaction is, in fact, achieved, the foundation will remain stable, even under the most adverse ambient conditions.

Several organic-based products are currently available to the industry which reportedly have a history of success. The treatment costs for these materials seem to vary from about $0.50 to $1.50 per sq ft of treated area. Development of sufficient field data could reduce this cost so as to become more than competitive with the use of lime. When a product is in development stages, the natural tendency is to overtreat to be safe. Gradually,

experience permits the applicator to withdraw to the most economical and optimum use, relative to the specific soil and locale.

CONCLUSION

The various repair procedures have been presented; however,
more often than not the actual field problem is not "cut and
dried," but requires expert analysis and usually some common
sense compromise. For example, consider a frame property with
a wood foundation resting essentially on grade (no crawl space).
Further complicate the problem by assuming rotted plates, joists
or girders, and a market value of the property at $8,000. To
raise the residence above grade to allow work space and installation of any reasonable supports would cost in the range of
at least half the market value of the property without commen-
surate increase in that market value. Obviously foundation
repairs would be difficult to justify in this instance. This
example is admittedly extreme but in older properties, even
with higher values, some related compromise is often advised.
Proper foundation repairs start with a practical as well as
technical evaluation of the situation including such influences
as the value of the property, cause and extent of distress,
approaches to repair, cost of repairs, value of repairs to structure, etc. Often the wisest approach is to level to a practical
extent and stabilize to arrest future movement. In this case,
the foundation may not be level but it is stable. This compromise
normally loses little with respect to serviceability or value of the
property. As a matter of fact, few, if any, new residential
foundations are constructed truly level. Even the most competent foundation repair specialists may occasionally make a
judgment error but even so, in the long run, the property
owner often benefits. The alternate is to overdesign the repairs
in an attempt to cover all possible contingencies, including
perhaps some ignorance regarding the problem itself. This

invariably results in overpricing. The author has seen competitive bids, on the same repairs, vary by a factor of four or more.

BIBLIOGRAPHY

1. Craft et al. *Well Design: Drilling and Production.* Englewood Cliffs, N.J.: Prentice-Hall, 1962.

2. Frederick, F. R. *Rheology.* New York: Academic Press Vols. 1–3, 1956.

3. O'Neill, M. W. and Nadar Poormooyed. "'Methodology for Foundations on Expansive Clays." *Journal of the Geotechnical Engineering Division,* ASCE, Vol. 106, No. GT12, Dec. 1980, pp. 1345–1365.

8

DETERRENTS TO PROPER LEVELING

This chapter will touch on those inherent problems which impede the intended foundation leveling or some related operation. In most cases, specific and marked improvement of the condition is experienced though the overall results might be less than desired. Factors to be considered will include upheaval, warped wood members, and various construction or "as-built" conditions.

UPHEAVAL

As noted in previous chapters upheaval is that condition where some area of the foundation is forced above the original grade. Upheaval is caused by water accumulation and subsequent swell of the clay soil which exerts an upward force on the foundation. When this condition occurs it is rare that the foundation can be completely restored to a level condition. As a rule the lowermost, surrounding areas are raised to approximate or "feather" the heaved crown, and the foundation is then stabilized to prevent future distress. This is normally acceptable as far as utilization is concerned.

WARPED SUBSTRUCTURE

Warped wood members are another antagonist to foundation leveling. This problem normally involves pier and beam construction but can also include slab construction where the floors are laid on a wood-screed system. In either event a wood member warps into an arc due to the externally applied stress caused by differential movement in the foundation. In the stressed condition the wood is in tension and thus often tight against the foundation supports. Upon raising the extremities (ends) of the wood in the leveling attempt the stress is relieved but the member retains the warped or distorted state (see Figure 8-1). The floor surfaces at points A and B are level after the raising operation but the midsection at C is above the desired grade. From a foundation repair standpoint nothing further can be done to improve the situation. Often, given time, the wood will "relax" and the crown will depress to approximate normal. In an effort to encourage this event the appropriate (hopefully) space is left above the shim at point C. In cases of slab construction with wood screed, the analogy is identical except that the screed is warped rather than the girder.

CONSTRUCTION INTERFERENCES

Often specific conditions of construction will restrict or prevent the desired foundation repair results. A few specific examples will be discussed in the following paragraphs.

Add-On or Remodeling Without Correcting Preexisting Foundation Distress

Often the condition exists where an add-on is constructed abutting an existing structure which has previous and ongoing differential movement. In time the condition progresses to the extent that foundation leveling is desired. With a pier and beam the only technique is to underpin (mechanically raise) the affected beam section. To accomplish this, proper access must be provided. Generally this would require tearing out floor sections in either the add-on or the original structure and doing the work from within. Obviously this would be quite expensive. For slab construction the problem area might be restored to intermediate grade by mud-jacking inside the original structure. Generally this is not a serious problem and results are satisfactory. In either event, slab or pier and beam, the original

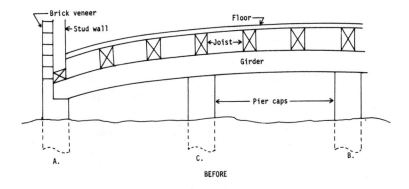

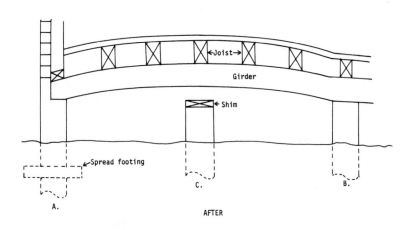

Figure 8-1. Warped substructures.

structure is restored to that grade or level existing when the add-on was added and not necessarily to level. To attain a nearly true level would require raising the entire add-on to the proper established grade consistent for the original structure. Again this approach would involve considerable expense and generally the results would not justify or warrant the cost.

Remodeling operations can create a similar problem. Suppose a wall partition is added in an area where the floor is not level. If that partition is built rigid, one of two things will occur; either some wall studs will be longer than others or one or both plates will be shimmed. Leveling the floor will then push the "longer" stud area through the ceiling. If this condition is intolerable the leveling must be curtailed short of the desired results. (The "plates" are wood members placed horizontally at each end of the stud wall. The ceiling plate forms the top header and the floor or sole plate the bottom. See Figures 4-1 or 4-5 for examples).

Creation of New Distress

Most structures are not initially constructed perfectly level. Any attempt to correct an "as-built" problem will result in the creation of new distress. For this reason a foundation is normally raised to a "level appearance" rather than a "true level." Even then, some compromise is often required to equate the degree of levelness with practical job conditions. For example, if a leveling operation intended to close a ¼ in. crack develops an offsetting ½ in. crack, nothing is being gained and a decision or compromise must be made. Usually all attempts to raise are ceased.

The same is true in the instance of offsetting doors. Assume two doors are opening in opposite directions, either on 180 or 90 deg planes. If door A is plumb and door B is not, little or nothing constructive can be gained by attempting to level door B. Efforts to improve door B will most often cause an equivalent, offsetting movement to door A.

Occasionally an upper floor will develop a sag in an area not directly supported by the lower floor or foundation. Since there is no vertical connection between the problem area and the lower floor, foundation leveling will neither affect nor improve the sag. One solution has been to install wood beams across the lower floor ceiling, properly anchored, to raise and support the area. Alternately, where practical, columns might be installed to provide the support and/or to provide contact with the foundation. This latter solution is often not acceptable — particularly with residential construction. The columns would be obstructive to movement and generally unsightly.

Incompentent Foundations

The competency of the foundation materials are sometimes below standard and, hence, prevent or deter proper foundation repairs. Generally, this condition involves older concrete foundations either poured with faulty concrete or deteriorated by unusual chemical or weather attack. In other instances the problem might involve rotted or deficient wood. If the foundation will not withstand the stress required for leveling, little can be accomplished. In some marginal conditions, long steel plates may be placed under the foundation member to distribute the load and thus facilitate leveling. This process is relatively expensive and does not replace the deficient materials. Faulty wood can generally be replaced economically if proper access or work space is available. Replacing faulty concrete is generally quite expensive and, since older properties are most often involved, not economically feasible.

Interior Fireplaces

When interior fireplaces settle, expensive problems also develop. Even with slab foundations the usual repair technique requires the installation of spread footings (typical) and mechanical raising. Access then becomes a serious and costly obstacle. If the movement is not substantial, an acceptable approach is to either cut the surrounding areas loose from the fireplace and level the floors by normal shimming operation (pier and beam foundations) or, in the case of slabs, stabilize the foundation to, hopefully, prevent future, progressive movement.

Crossbeams, Interior Piers, or Intensive Loads (Slabs)

In slab construction, the condition sometimes develops where a problem area involves a stiffened slab (crossbeams) or intensified load offset by a normal slab floor section. Attempts to raise the weighted area by mud-jacking would be susceptible to crowning the weaker slab. In this instance hydraulic leveling attempts (mud-jacking) must be aborted to protect the remaining slab. The alternative would be to break out the adjacent slab for access and mechanically underpin the problem area. As a rule this is not advisable both from a cost and structural view. The identical problem is encountered in cases where interior piers (belled or otherwise) are tied into the slab. Again the general selection would be to level to a practical extent and stabilize to prevent future, progressive movement.

Lateral Movement

Under certain conditions one foundation member may move laterally with respect to another. Specific examples would be a patio slab moving away from the perimeter beam or a perimeter beam moving horizontally from an interior floor. In either instance the problem member can be raised vertically quite easily; however, it is altogether another problem to move that structure laterally to restore the original position. As a rule this is beyond reasonable expectation. Prevention is therefore the best solution.

There are other conditions which interfere with proper or desired foundation leveling; however, those presented on the preceding pages are the most common.

9

PREVENTIVE MEASURES

PROPER MAINTENANCE

Preventing a problem is always more desirable than having to cure one. Certain maintenance procedures can help prevent or arrest foundation problems if initiated at the proper time and carried out diligently. The following are specific suggestions which help encourage foundation stability.

WATERING

In dry periods, summer or winter, water the soil adjacent to the foundation to help maintain a constant moisture. *Proper* watering is the key.

When cracks appear between the soil and foundation, the soil moisture is low and watering is in order. On the other hand, water should not be allowed to stand in pools against the foundation. Watering should be uniform and preferably should cover long areas at each setting, ideally 50 to 100 lin ft. Too little moisture causes the soil to shrink and the foundation to settle. Too much water — an excessive moisture differential — can cause the soil to swell and heave the foundation. Along these lines never attempt to water the foundation with a root feeder or by placing a running garden hose adjacent to the beam. Sprinkler systems often create a sense of "false security" because the shrub heads, normally in close proximity to the perimeter beam, are set to spray away from the structure. The design can be altered to put water at the perimeter and thereby serve the purpose quite adequately. The use of a soaker hose is normally the best solution. From previous studies of infiltration and runoff it became evident that watering must be close to the foundation, within 6 to 18 in., and excessive watering can be prevented by proper grade around the foundation.

A more sophisticated watering system is now available utilizing a subsurface weep hose with electric activated control valves and automatic moisture monitoring and control devices. A typical system is outlined in Figure 9-1 and Figure 9-2. The multiple moisture control devices are situated to afford adequate soil moisture control automatically and evenly around the foundation perimeter. Reportedly, the moisture control can be set to maintain effective soil moisture variations to plus or minus 1%. Within this tolerance little, if any, differential foundation move-

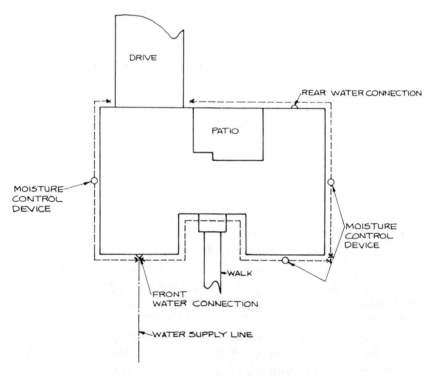

Figure 9-1. Schematic diagram for watering system around perimeter beam.

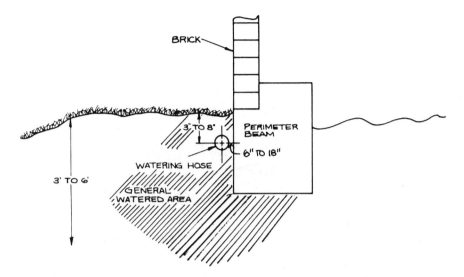

Figure 9-2. Typical location of watering hose with respect to both depth and proximity to perimeter beam.

ment would be expected in even the most volatile or expansive clay soils.

DRAINAGE

For the reasons noted above, it is important that the ground surface drain away from the foundation. Where grade improvement is required the fill should be a low or clay-free soil. The slope of the fill need not be exaggerated but merely sufficient to cause the water to flow outward from the structure (see Figure 9-3). The surface of fill must be below the air vent for pier and beam foundations and below the brick ledge for slabs. Surface water, whether from rainfall or watering, should never be allowed to collect and stand in areas adjacent to the foundation wall. Consistent with this, guttering and proper discharge of downspouts is quite important. Flower bed curbing and planter boxes should drain freely and preclude trapped water at the perimeter. In essence, any procedure which controls and extricates excess surface water is beneficial to foundation stability.

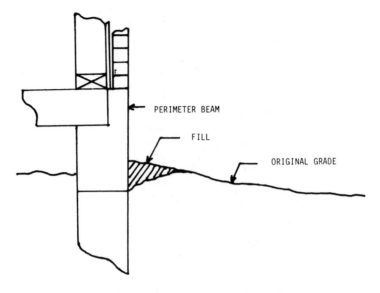

PERIMETER BEAM

FILL

ORIGINAL GRADE

Figure 9-3. Correct drainage (typical).

VEGETATION

Certain trees, such as the weeping willow, grow extensive shallow root systems. These plants can cause foundation (and sewer) problems even if located some distance from the structure.

Many other plants and trees can cause foundation problems if planted too close to the foundation. Plants with large, shallow root systems can grow under the foundation and, as roots grow in diameter, produce an upheaval in the foundation beam. The FHA suggests that trees be planted no closer than their ultimate height; a safer distance should be at least 1.5 times the ultimate anticipated heighth.

Other plants remove water from the foundation soil (transpiration) causing a drying effect which in turn could produce foundation settlement.

Any extended differential in soil moisture can produce a corresponding movement in the foundation. If the differential movement is extensive, foundation failure will likely result. Of the two "categorized" type failures, settlement and upheaval, the latter is by far the more critical.

Even with proper care, foundation problems can develop; however, consideration and implementation of the foregoing procedures will afford a large measure of protection. It is possible that adherence to proper maintenance could eliminate perhaps 40% of all serious foundation problems.

10

FOUNDATION INSPECTION AND EVALUATION FOR THE RESIDENTIAL BUYER

If you are house shopping in a geographical locale with a known propensity for differential foundation movement, it is always wise to engage the assistance of a qualified foundation inspection service. The word qualified cannot be overemphasized. One should both evaluate the existence of foundation-related problems and also determine the *cause* of the problems when such are found to exist. The latter is particularly important! Repairs will be futile if the original cause of the distress is not recognized and eliminated.

Figure 6-1 (A, B, C, and D) depicts several of the more obvious manifestations of distress relative to differential foundation movement. With only limited experience one can learn to detect these signs. Often the problems of detection become more difficult when cosmetic attempts have been used to conceal the evidence. These activities commonly involve painting, patching, tuck-pointing, addition of trim, installation of wall cover, and so forth, and, as one would guess, require greater expertise for evaluation.

The important issue in all cases is to decide whether the degree of distress is sufficient to demand foundation repair. This decision requires a great degree of experience since several factors influence the judgment such as:

(1) the extent of vertical and lateral deflection (does any structural threat exist or appear imminent?), (2) whether the stress is ongoing or arrested, (3) the age of the property, (4) the likelihood that the initiation of adequate maintenance would arrest continued movement (for example, in cases of upheaval, elimination of the source for water will often arrest the movement), (5) value of the property as compared to repair costs (most foundation repair procedures require some degree of compromise). In order to arrive at a reasonable or practical cost, the usual primary concern is to render the foundation "stable" and the appearance "tolerable." In most cases, the cost to truly "level" a foundation, if possible, would be prohibitive.). (6) age, type, and condition of the existing foundation, and (7) the possibility that, if the movement appears arrested, cosmetic approaches would produce an acceptable appearance.

It is difficult, if not impossible, to properly evaluate the above without extensive first-hand experience with actual foundation repairs. One good example of the importance of on-the-job experience is the proper determination of upheaval as opposed to settlement. (Refer to Chapter 5 and Figure 5-3.) If upheaval were evaluated as settlement, the existence of water beneath the foundation would likely be overlooked. In which case future, more serious, consequences would be nearly certain. However, there are no similar, serious consequences in making the error of labeling settlement as upheaval. Essentially, this is true because all foundation repair techniques are designed to raise a lowermost area to some higher elevation. In cases of settlement, one merely raises the distressed area to "as built." In instances of upheaval, it becomes necessary to raise the "as built" to approach the elevation of the distorted area. The latter is obviously far more difficult. (Refer to Chapter 5 for greater detail.)

The existence of foundation problems need not be particularly distressing so long as the buyer is aware of potential problems

Table 10-1

Check List For Foundation Inspection

A. Check the exterior foundation and masonry surfaces for cracks, evidence of patching, irregularities in siding lines or brick mortar joints, separation of brick veneer from window and door frames, trim added along door jam or window frames, separations or gaps in cornice trim, separation of brick from frieze or fascia trim (look for original paint lines on brick), separation of chimney from outside wall, etc.

B. Sight ridge rafter, roof line, and eaves for irregularities.

C. Check interior doors for fit and operation. Check for evidence of prior repairs and adjustment such as shims behind hinges, latches or keepers relocated, tops of doors shaved, etc.

D. Check the plumb and square of door and window frames.

E. Note grade of floors. A simple method for checking the level of a floor (without carpet), window sill, counter top, etc., is to place a marble or small ball bearing on the surface and observe its behavior. A rolling action would indicate a "down hill" grade. (A hard surface such as a board or book placed on the floor will allow the test even on carpet.)

F. Inspect wall and ceiling surfaces for cracks or evidence of patching. *Note:* any cracking should be evaluated on the basis of both extent and cause. Most hard construction surfaces tend to crack. Often this can be the result of thermal or moisture changes and not foundation movement. However, if the cracks approach or exceed 1/4 in. in width, the problem is likely to be structural. On the other hand, if a crack is noticed which is, say, 1/16 in. wide, is it a sign of impending problems? A simple check to determine if a crack is "growing" is to scribe a pencil mark at the apex of the existing crack and using a straight edge, scribe two marks along the crack, one horizontal and one vertical. (Refer to Figure 10-1.) If the crack changes even slightly one or more of the marks will no longer match in a straight line across the crack and/or the crack will extend past the apex mark. (Refer to Figure 10-2.) A slight variation of this technique is to also scribe a straight-line mark across a door and matching door frame. In a structure older than about 12 months, continued growth of the crack would be a strong indication of foundation movement.

G. On pier and beam foundations, check floors for firmness, inspect the crawl space for evidence of deficient framing or support, and ascertain adequate ventilation.

H. Check exterior drainage adjacent to foundation beams. Any surface water should quickly drain away from the foundation and not pond or pool within 8 or 10 ft distance. Give attention to planter boxes, flower bed curbing, and down spouts on gutter systems.

I. Look for trees that might be located too close to the foundation. Most authorities feel that the safe planting distance away from the foundation is 1, or preferably 1.5, times the anticipated ultimate height of the tree. Consideration should be given to the type of tree.

J. Are exposed concrete surfaces cracked? Hairline cracks can be expected in areas with expansive soils. However, larger cracks approaching or exceeding 1/8 in. in width warrant closer consideration.

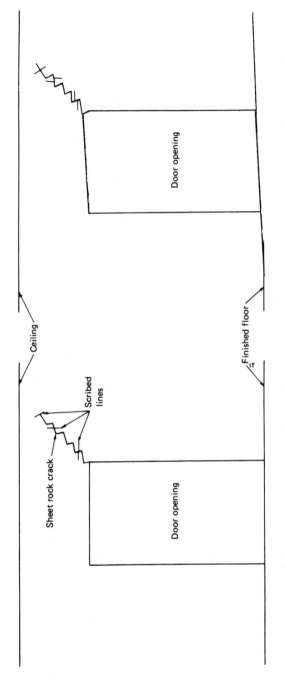

Figure 10-2. Movement is confirmed by changes in planes of scribed lines and extension of crack apex.

Door opening

Ceiling

Finished floor

Scribed lines

Sheet rock crack

Door opening

Figure 10-1. Scribe cracks to monitor suspected movement.

before the purchase. As a rule, the costs for foundation repair are not comparatively excessive, and the results are most often satisfactory. It is the surprise that a buyer cannot afford. The cost for the residential foundation inspection service is quite nominal, varying (in 1983) from about $70.00 to $150.00, depending on the time involved and the locale. Table 10-1 presents a simple check list for evaluating the stability of a foundation. If you find questions or uncertainty regarding any of the items, consult a qualified authority immediately.

GLOSSARY

Adequate watering: will help prevent or arrest settlement brought about by soil shrinkage resulting from loss of moisture.

Aeolian: soil deposition by wind.

Aeration zone: Includes the capillary fringe, intermediate belt (which may include one or more perched water zones) and, at the surface, the soil water belt, often referred to as the root zone.

Alluvial: soil deposition by river waters.

Arenaceous: sandy soil.

Argillaceous: clayey soil.

Basal spacing: the distance between individual or molecular layers of the clay particles.

Caliche: argillaceous limestone or calcareous clay.

Capillary fringe: contains capillary water originating from the water table.

Capillary rise: is impeded by the swell of clay particles (loss of permeability) upon invasion of water. Finer soils will create a greater height of capillary rise but the rate of rise is slower. A measure of height of water rise above the level of free water boundary.

Clays: these are the finest possible particles usually smaller than 1/10,000 of an inch in diameter.

Cohesion: a cementing or gluing force between particles; requires a clay content.

Dead load: the weight of the empty structure.

Field capacity: residual amount of water held in the soil after excess gravitational water has drained and after the overall rate of downward water movement has decreased (zero capillarity).

Fill: site is not level and requires filling with soil.

Footing design: is to distribute the foundation load over an extended area and thus provide increased support capacity on any substandard bearing soil.

Foundation: is that part of a structure in direct contact with the ground which transmits the load of the structure to the ground.

French drain: is installed to intercept and divert the water; cut to depth below the level of the intruding water.

Frost heaving: occurs when a mixture of soil and water freezes. When it freezes the total volume may increase by as much as 25% dependent upon the formation of ice lenses at the boundary between the frozen and unfrozen soil.

γ_T : is unit weight of liquid at temp. T, gm/cc.

Grout curtain: a continuous consolidated boundary or area of sufficient strength to permit excavation or provide adequate bearing strength.

Grouting operations: include an activity whereby the success of the operation depends upon the ability of an injected material to penetrate and permeate a relatively deep soil bed.

Gumbo: highly plastic clay from the southern and/or western U.S.A.

Gypsum: hydrous calcium sulfate.

Interior floors: are supported by piers and pier caps which support the girder and joist system of a wood sub-structure.

Interlayer moisture: is that water situated within the crystalline layers of the clay. This water provides the bulk of the residual moisture contained within the intermediate belt.

Intermediate belt: contains moisture essentially in dead storage — held by molecular forces and may include one or more perched water zones.

Live load: is the weight of building contents, plus wind, snow, and earthquake forces where applicable.

Loess: an aeolian deposit of uniform gradation with some calcareous cementation.

Low profile: where the crawl space is substantially lower than the exterior grade.

Marl: a calcareous clay.

Maximum density: as water is added, the first volume fills voids and helps the particles move closer together thus increasing the density. Extra water beyond the optimum displaces the heavier solids and thus reduces the density, hence the "optimum" water produces the maximum density.

Moisture barrier: a remedial technique for maintaining moisture content beneath a foundation (generally slab) utilizing an impermeable barrier extending to some depth and in close proximity to the perimeter beam.

Mud-jacking: is a process whereby a water and soil cement or soil-lime-cement grout is pumped beneath the slab, under pressure, to produce a lifting force which literally floats the slab to desired position.

Newtonian: fluids (e.g., water, antifreeze, salt water, etc.)that produce a straight line through the origin when a shear rate - shear stress diagram is plotted.

Non-cohesive: there is no attraction or adhesion between individual soil particles.

Non-Newtonian: fluids (e.g., cement, soil or soil cement slurries) that do not exhibit a linear shear rate - shear stress diagram.

Normal design: wherein the crawl space is at a grade equivalent to the exterior landscape.

Perched water zones: the perched ground water if it occurs, develops essentially from water accumulation either above a relatively impermeable strata or within an unusually permeable lens, it generally occurs after a good rain and is relatively "temporary".

Perma-jack: utilizes a hydraulic ram to drive jointed sections of $3''$ steel pipe to rock or suitable bearing.

Phreatic boundary: surface of the water table which will not normally deflect or deform except under certain conditions in the proximity of producing well.

pH: a measure of the available hydrogen ion concentration.

Pier and beam: design wherein the perimeter loads are carried on a continuous beam supported inturn on piers drilled into the ground, supposedly to a competent bearing soil or strata.

Piers or pilings: are normally extended through the marginal soils to either rock or other competent bearing material.

Pipe flow apparatus: represents the simplest capillary viscometer and is, as expected, the most commonly used.

Plasticity index (PI): is a dimensionless constant which bears a direct ratio to the affinity of the bearing soil for volumetric changes with respect to moisture variations. The PI is determined as the difference between the liquid limit (LL) and plastic limit (PL).

Poorly graded: a majority of soil particles are of one particular size in the soil.

Pore water: occur within the soil mass external to individual soil grains, and is held by interfacial tension.

q_u: the unconfined compressive strength which measures its capacity to carry the load.

r: is the radius of pore (capillary) in cm:

Rock: is not always a superior foundation bed, depending upon such factors as the presence of bedding planes, faults, joints, weathering, cementation of constituents, etc.

Root zone: Upper layer area where plant roots take their moisture out of soil.

Run off: the excess water not retained by the soil.

Sands and gravels: are the easiest soil types to distinguish. These consist of coarse particles which range in size of from 3″ in diameter down to small grains which can be barely distinguished by the unaided eye as separate grains.

Saturation zone: is more commonly termed the "water table" or ground water, and is, the deepest soil water source.

Settlement: is that instance in which some portion of the foundation drops below original "as built" grade.

Shale: sedimentary rock; indurated clay and/or silt muds.

Silts: are finer than sands but more coarse than clays. They represent soil particles of ground rocks which have not yet changed their character into the minerals.

Slab: construction of one variety or another wherein the concrete foundation is supported entirely by surface soils and which probably constitutes the majority of new construction in geographic areas exposed to high-clay soils.

Sliding: occurs when a structure is errected on a slope—herein the movement is not limited to up or down (vertical) but possesses a lateral or horizontal component.

Soil: generally describes all the loose material constituting the earths' crust in varying proportions, and includes three basic materials: air, water, and solid particles. The solid particles have been formed by the disintegration of different rocks.

Soil belt: can contain capillary water available from rains or watering; unless this moisture is continually restored the soil will eventually desiccate through the effects of gravity, transpiration and/or evaporation.

Soil water belt: provides moisture for the vegetable and plant kingdom.

Supports or spread footings: (1) steel-reinforced footings of sufficient size to adequately distribute the beam load over the supporting soil and poured at a depth to be relatively independent of seasonal soil moisture variation (2) a steel-reinforced pier tied into the footing with steel and poured to the bottom of the foundation beam.

Surface absorbed water: occur within the soil mass external to individual soil grains and is held by molecular attraction between the clay particle and the dipolar water molecule.

Transpiration: refers to the removal of soil moisture by vegetation.

T_{ST}: is surface tension of liquid at temp. T

Upheaval: relates to the situation where the internal areas of the foundation raise above the as built position.

Viscosity: a single constant that describes the Newtonian relationship of shear stress and shear rate.

Void ratio: the ratio of combined volume of water and air to the total volume of the soil sample.

W: is the sum of live loads (^{W}L) and dead loads (^{W}d).

Water leaks: are accumulated under the slab, and any water under the slab regardless of source, tends to accumulate in the plumbing ditch.

Well graded: a fairly even distribution of grain sizes in the soil.

INDEX

INDEX

111